# SpringerBriefs in Computer Science

**Series Editors**

Stan Zdonik, Brown University, Providence, RI, USA
Shashi Shekhar, University of Minnesota, Minneapolis, MN, USA
Xindong Wu, University of Vermont, Burlington, VT, USA
Lakhmi C. Jain, University of South Australia, Adelaide, SA, Australia
David Padua, University of Illinois Urbana-Champaign, Urbana, IL, USA
Xuemin Sherman Shen, University of Waterloo, Waterloo, ON, Canada
Borko Furht, Florida Atlantic University, Boca Raton, FL, USA
V. S. Subrahmanian, Department of Computer Science, University of Maryland
College Park, MD, USA
Martial Hebert, Carnegie Mellon University, Pittsburgh, PA, USA
Katsushi Ikeuchi, Meguro-ku, University of Tokyo, Tokyo, Japan
Bruno Siciliano, Dipartimento di Ingegneria Elettrica e delle Tecnologie
dell'Informazione, Università di Napoli Federico II, Napoli, Italy
Sushil Jajodia, George Mason University, Fairfax, VA, USA
Newton Lee, Institute for Education, Research, and Scholarships,
Los Angeles, CA, USA

SpringerBriefs present concise summaries of cutting-edge research and practical applications across a wide spectrum of fields. Featuring compact volumes of 50 to 125 pages, the series covers a range of content from professional to academic.

Typical topics might include:

- A timely report of state-of-the art analytical techniques
- A bridge between new research results, as published in journal articles, and a contextual literature review
- A snapshot of a hot or emerging topic
- An in-depth case study or clinical example
- A presentation of core concepts that students must understand in order to make independent contributions

Briefs allow authors to present their ideas and readers to absorb them with minimal time investment. Briefs will be published as part of Springer's eBook collection, with millions of users worldwide. In addition, Briefs will be available for individual print and electronic purchase. Briefs are characterized by fast, global electronic dissemination, standard publishing contracts, easy-to-use manuscript preparation and formatting guidelines, and expedited production schedules. We aim for publication 8–12 weeks after acceptance. Both solicited and unsolicited manuscripts are considered for publication in this series.

More information about this series at http://www.springer.com/series/10028

Mario Alemi

# The Amazing Journey of Reason

from DNA to Artificial Intelligence

 Springer Open

Mario Alemi
Elegans Foundation
London, UK

ISSN 2191-5768          ISSN 2191-5776   (electronic)
SpringerBriefs in Computer Science
ISBN 978-3-030-25961-7          ISBN 978-3-030-25962-4    (eBook)
https://doi.org/10.1007/978-3-030-25962-4

This Springer imprint is published by the registered company Springer Nature Switzerland AG
The registered company address is: Gewerbestrasse 11, 6330 Cham, Switzerland

# Preface

In the park, in Milan, where I am writing this preface, there are seven other people enjoying "nature". Four of us are sitting on benches, either using a smartphone or a laptop (myself). Two are on the grass, talking. The remaining two, a boy and a girl, jokingly fight in the area with trees. The boy initially pretends to lose and, then, to prove his athletic abilities, climbs on a small tree imitating a chimpanzee.

In a few square metres, millions of years of evolution of our species are represented – the arboreal past, with grooming, the spoken language in the savannah, and finally digital communication.[1]

I cannot avoid asking myself, once more – How did we get here, and where are we going? With this question in mind, I started looking at the development of life inside a single coherent framework – theory of information and theory of network – with the hope to shed some light on the possible futures humanity could face.

There is no definitive answer, but scientific research has proposed a few theories which, put together, present an interesting picture. And I am not just talking about paleoanthropology. We are children of the universe, and the laws governing the universe are the same as those which enabled first the emergence of life, then of *Homo sapiens*, and finally of human societies – artificial intelligence included.

One of the characteristics of life is that it becomes increasingly more complex. At the same time, life devises increasingly more sophisticated tools with which to gather energy from the environment. Yet another characteristic of life is self-similarity: a human society has a similar structure to our own person (with a government/brain, working force/muscular system, army/immune system, etc.), which in turn has a similar structure to a unicellular organism.

This self-similarity is an effect of complexity being built by successive aggregations: atoms into molecules, molecules into amino acids, and these into proteins, then cells, then complex organisms, and finally societies. Every form of life is made

---

[1] There are also people talking to dogs and collecting their excrement, a behaviour which I still find hard to justify from any evolutionary point of view.

up of less complex elements, collaborating so tightly that, as Jean Jacques Rousseau said, each element alienates itself, totally, to the community.

There is a moment when the community becomes a form of life on its own. Our own body, a community of collaborating cells, is a living being, not just an aggregation of cells. The whole is much more than the sum of its parts: what our cells can do together cannot be compared to what they would do independently.

This is the mantra of the book – to evolve, life needs collaboration. I am not preaching a religion of love. I am saying that the physics of life is based on collaboration. Life does not work without the ability of individuals to collaborate.

The necessity to collaborate comes from the fact that living systems are systems which process information on how to extract energy from the environment. They need this energy, because processing information requires energy – think of your brain or your smartphone. The more information processed, the more energy extracted, but also the more energy needed.

Aggregation happens because, sooner or later, organisms reach a limit in terms of how much information they can process. Biological cells, at a certain point, could not get any bigger. They therefore started collaborating, forming complex organisms. A few billion years later, one of these complex organisms, *Homo sapiens*, built a very powerful brain that also reached a limit – so it started building a metaorganism made up of many collaborating *Homo sapiens*. These metaorganisms were based initially on language, then writing and math, then printing, and now digital communication and data processing.

And so we end up with artificial intelligence and the emergence of the distributed nervous system of a metaorganism, in which human beings are the cells, a reassuring world but one where each individual will count increasingly less.

A neuron's life is safe, but not fun: little more than receiving inputs and generating outputs. Neurons possess no real knowledge about the external environment. They don't even "know" they are part of the brain. A neuron, left alone, dies. An amoeba, which we consider a primitive form of life, survives, because it knows the environment it lives in.

Similarly, we *sapiens* are becoming information processing apes – good at communicating through our new digital global nervous system but increasingly incapable of storing and processing information in our own brain. We are experiencing the highest survival rate of our history but are increasingly less knowledgeable about the world we live in: the information is now processed by the network of people and machines, not by the individual. Like neurons, it is becoming difficult for us to understand how our society works and what the forces are moving it. And like neurons, we would not survive for long outside our society, contrary to so-called primitive people, who still retain lots of information about the environment.

Currently, in this landscape, we have organisations, whose only mission is increasing their annual revenues – *their* energy input – managing such global nervous system. As if that was not enough, our energy consumption has become unsustainable: per capita consumption is around 5,000 times that of early hominids, and it is based on non-renewable resources.

Are we doomed? In the long term, any form of life is. But in the medium short term, it is possible that the human metaorganism will not only survive but also thrive. The emergence of the DNA in cells or the brain in animals was pivotal for the evolution of life, and it is possible that artificial intelligence will play a similar role.

London, UK                                                                        Mario Alemi

# Skipping Math

The math and physics in the book are kept at a minimum level. A few sections would be easy to follow for a "hard" scientist, less for someone who does not like formulas. I believe the book can be read skipping those parts, and in order to make the process easier, a few sections which I believe can be skipped are in *italics*.

For example, understanding the difference in complexity between the numbers "$\pi$" and "1" is not fundamental for reading the book, but it helps understand what complexity is. For that reason, it is still in the book but in *italics*.

"Cult anthropology is that branch of natural science which deals with matter and motion, *i.e. energy*, phenomena in cultural form, as biology deals with them in cellular, and physics in atomic, form."

Leslie Alvin White (1943). *Energy and the evolution of culture*. American Anthropologist.

...from schools to universities to research institutes, we teach about origins in disconnected fragments. We seem incapable of offering a unified account of how things came to be the way they are.

David Christian (2004). *Maps of time. Introduction to Big History*. University of California Press.

[My father] was reading books on the brain, looking for clues about how to make a computer intuitive, able to complete connections as the brain did ... the idea stayed with me that computers could become much more powerful if they could be programmed to link otherwise unconnected information.

Tim Berners-Lee (1999). Weaving the Web: The original design and ultimate destiny of the World Wide Web by its inventor. DIANE Publishing Company.

# Acknowledgements

I always found it slightly cheesy when the authors started by acknowledging the importance of their family in bringing their work to light. Cheesy or not, I could not dedicate this book to anyone other than my parents. Without their passion for knowledge – we had hundreds of books when books were the main access to knowledge – I may never have become a scientist.

They were no scientists but enjoyed my obsession with science. I once explained to my mother the expensive tissue hypothesis (you will read about that later). It was a few months before she passed away, due to a terminal illness. She thought for a few seconds over what I said and then said, simply, with a smile, "It must be really nice to see things the way you see them".

My father, who suddenly passed away during the writing of this book, was extremely excited about it. He wanted constant updates and even organised a conference where I could test the topics in the book – in front of an audience of extremely inquisitive members of the local Rotary Club.

Similarly, my sister, Francesca, and her family – Carolina, Riccardo, Federico, and Pietro – showed immense kindness in their support.

If my family made possible, with their love and care, the genesis of the book, the real "authors" of this book are the hundreds of people cited in the bibliography. Any originality in this work is not in the ideas presented but in the way these ideas are connected. Along the spirit of the book, I hope I created a network with an interesting emergent property.

Still, I don't feel at ease citing intellectual giants like John von Neumann, Herbert Simon, or Lynn Margulis. There is always the doubt that my interpretation of their thoughts is correct. This doubt is even worse when I openly say that I don't agree with giants of popular science like Richard Dawkins or Yuval Harari. I firmly believe that every single author I quote deserves my gratitude – independently of whether I agree with them or not.

That said, I could never have connected these ideas without the people who make my life so worth living – my friends. While at CERN[1] in 1996, I tried to (impolitely) ridicule, in front of others, Adolfo Zilli's idea that energy and information are connected. He was right, and 10 years later, he became the sounding board who helped transform a few confused ideas into a hopefully more coherent framework. Together with his partner and my dear friend Elisa Cargnel, we have spent countless evenings discussing the topic.

Diana Omigie gave me the first books I read on network theory and has been a fantastic person to talk to about network and information: as a (then) young neuroscientist, she was very happy to discuss those topics and always encouraged me to go further.

During a dinner in Milan in 2012, Toni Oyry and Lina Daouk convinced me I had to write a pitch for a literary agent. Their enthusiasm, which translated into various trips to Lebanon, in order to outline the contents of the book, was a key factor in the genesis of the project.

Cristina Miele, my partner for many years, kindly dedicated numerous weekends and holidays to (my) writing. If this book was a movie, she would have been the executive producer. Similarly, my business partner and dear friend Angelo Leto never complained if I stole a few hours from work because I felt the urge to write. On the contrary, he was one of the first friends to read the book.

Endless people were victims of my obsession and were glad to discuss the ideas you'll find exposed here. Sometimes, they were friends. Sometimes, they were people I barely knew, who, after showing some initial interest, found themselves locked in endless discussions.

During a trip to Japan in 2013, I met a Neapolitan salesman, Marco Senatore, with whom I discussed, the whole night in an enchanted Japanese garden, the book I wanted to write. This developed into an unexpected friendship, still alive today. Soon after, Tommaso Morselli (the first person to read the final version of the book) in Bologna was a great companion for drinking wine in the city's *osterias* while talking about the book. In 2014, while consulting at Mondadori, an Italian publisher, I met Matteo Spreafico. Again, endless nights with him have contributed to what the book became.

Similarly, there were fruitful discussions or even just a single long discussion with friends, who –sometimes without even knowing it – contributed to the final picture: Matteo Berlucchi, Giovanni Scarso Borioli, Chiara Ambrosino, Amana Khan, Tito Bellunato, Tommaso Tabarelli, Giovanni Caggiano, Stefania Sacco, Alessandra Tessari, Danilo Ruggiero, Ignazio Morello, and Teemu Kinos. In addition to comments, Jo Macdonald also provided the translation for most of the book.

Last but not least, I'd like to thank Yair Neuman. Three years ago, he read a post I wrote for my blog on the entropy of graphs. He found it interesting enough to contact me, introduce me to Springer Nature, and – as if this was not enough – read and edit the manuscript. Thanks to him, and Springer Nature's editor Susan Evan, this book is now in your hands – or e-reader.

---

[1] European Organization for Nuclear Research, Geneva, Switzerland

# Contents

# Chapter 1
# Life, Energy and Information

The principles we need to take into consideration while studying the evolution of how matter aggregates are quite simple.

To extract energy from a system, we need information about that system. We need to be able to predict how it will react, evolve. But to process information, in a brain, in a computer, in the DNA, we need energy.

Life means storing information to extract energy, and extract energy to store information. This chapter will analyse the concepts of energy and information, and how they relate to each other.

## What Is Life?

*Know thyself* is a good starting point for someone who wants to study its origin. And in our case, we must start with the question: what is life, what does it mean to be alive?

Erwin Schrödinger's, one of the parents of quantum mechanics, gave this definition in his booklet *What is life* (1944):

> Life is organized matter which evades the decay to equilibrium through absorption of energy

If we do not eat, all our cells, and then molecules, and the atoms, will be scattered in the environment. If we eat, we keep our body organised and avoid that.

A definition does not explain why life has emerged, or how it evolves. But is a good starting point to investigate these questions. Schrödinger definition allows us to clearly define the object of our interest, and is therefore worth to further explore it.

© The Author(s) 2020
M. Alemi, *The Amazing Journey of Reason*, SpringerBriefs in Computer Science, https://doi.org/10.1007/978-3-030-25962-4_1

## The Decay Towards Equilibrium

The concept of equilibrium mostly derives from the work of one single physicist: Ludwig Boltzmann. Towards the end of the nineteenth century, on the basis of the work done by Maxwell and others, Boltzmann introduced probability into the explanation of the evolution of systems into states of equilibrium.

The question no one could answer was –why there is such a thing as the equilibrium? Why, if I have a box filled with gas, I will never observe all molecules of gas in a single corner, but instead will see them always filling the whole box? Why do I need energy to compress all molecules in a corner?

The question is not much different from –why do I need energy to keep all molecules of an organism together? Why if I don't provide energy to the organism, all its molecules and atoms will eventually diffuse in the environment, like the molecules of gas in the box?

We can better comprehend Boltzmann's reasoning by simply imagining a gas with two molecules, one blue and one read, in a container with a semi-divider wall, as in Fig. 1.1.

If the two balls are not coloured, the states B′ and B″, with a ball on both sides, appear to be the same, so an external observer might call both states "macrostate B". If we shake the box, the macrostate B is more probable than L and R, because it is actually made of two *microstates*.

If there are one million balls (as many as are molecules in a cube of air with a 0.1 mm edge) the probability of obtaining states with approximately half the balls in each region is about $10^{300,000}$ times higher than that of obtaining states in which all the balls are on one side.

This is why we say that the system's equilibrium is with the balls, or molecules of gas, evenly distributed in the box. Unless we do something, the system will spontaneously evolve with the balls on both sides. Not always the same balls, but as far as we can say, the balls are evenly distributed.

**Fig. 1.1** A box, a wall and two coloured balls. There are four microstates, i.e. states which we can identify as different thanks to the color of the balls. When the balls are indistinguishable, like molecules in a gas, we only identify three macrostates

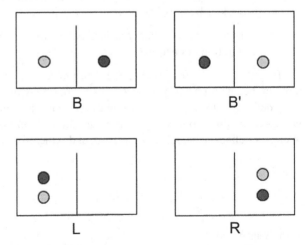

Similarly, if living systems do not absorb energy, they decay towards equilibrium in the sense that all their molecules tend to occupy the most probable macrostate, which is the one with all molecules diffused, like the ones in the gas. The gas does that quickly, our molecules slowly, but the process is the same.

To avoid the decay to equilibrium, we need energy. With some effort, we can move the balls from one side and put them on the other side, so to remain in a non-equilibrium state. Similarly, with some energy we keep our organism… organised, and all its molecules together.

## Energy Extraction Requires Information

The real breakthrough for Boltzmann was linking the concept of equilibrium to the one of *entropy*, a physical quantity related to order and the ability to extract energy from a system.[1] The more a system is far from equilibrium, the more energy we can extract. Let's see what it means.

Around 80 years after Boltzmann's studies on statistics and equilibrium, one of the most brilliant minds of the twentieth century, John von Neumann[2] (1956), linked entropy and information with that definition:

> Entropy … corresponds to the amount of (microscopic) information that is missing in the (macroscopic) description.[3]

Entropy, says von Neumann, is a lack of information about the system. We have microscopic information when we can identify each ball (e.g. with colour), macroscopic when not (all balls look the same). The less we can discern different microstates, and aggregate them into less informative macrostate, the higher the entropy.

Let us put this definition together with the one given by James Maxwell (1902), one of the founders of thermodynamics:

> The greater the original entropy, the smaller is the available energy of the body.

---

[1] Boltzmann defines his famous definition of entropy, engraved on his tombstone as a fitting epitaph: $S = k \cdot log\ W$, where $W$ is the number of microstates equivalent to the macrostate observed and $S$ is the dimensional constant, known as the Boltzmann constant.

[2] von Neumann, a Hungarian, one of the many Jews who fled from inhospitable Europe to America in the 1930s, was considered the most gifted man in the world by the likes of Enrico Fermi, who after working with him on a problem, said "I felt like the fly who sits on the plough and says 'we are ploughing'" (Schwartz 2017).

[3] The complete quote reads: "The closeness and the nature of the connection between information and entropy is inherent in L. Boltzmann's classical definition of entropy … as the logarithm of the "configuration number". The "configuration number" is the number of a priori equally probable states that are compatible with the macroscopic description of the state – i.e. it corresponds to the amount of (microscopic) information that is missing in the (macroscopic) description"

Maxwell says that low entropy means being able to extract energy. Von Neumann that low entropy means having information about the system. Therefore, when we have information about a system, we are able to extract energy from it.

If we think about that, it is quite obvious. If we want to extract wealth from the stock market, we need to study it. We need to be able to know how it evolves. If we want to extract energy from a liter of fuel, we need to know the laws of thermodynamics, so that we can build an engine.

In order to get energy, we, like any other living system, must have information about the environment. This allows us to absorb the energy which allows us to escape equilibrium.

Having defined life, we ended up with the idea that living organisms are systems which collect information about the environment. They use this information to extract energy from the environment and keep themselves in a state far from equilibrium. Before asking ourselves why they do so, we need to define information.

## Defining Information

If we want to study living systems, which store information on how to extract energy from the environment, we want to have a clear definition of what information is.

Acquiring information on a system means becoming able to predict how that system evolves with less uncertainty than we did before.

For those keen on a bit of mathematics, below we define uncertainty as a function of probability, and information as a function of uncertainty.

*To do this, all we have to do is define the level of surprise for an event. Surprise is a function of the probability* p, *where* p *indicates how strongly, in a 0–1 range, we believe that an event is going to happen. Surprise should therefore be big in the case of a small* p *(we are very surprised if something we think as improbable happens) to zero in the case of* p = 1 *(we are not surprised if something we consider to be inevitable happens).*

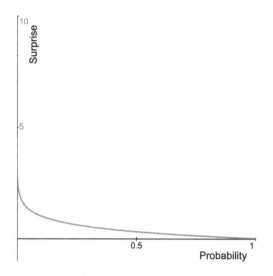

For those interested in a bit of math, the function that satisfies this relationship between p and surprise is the logarithm. As in Shannon (1948), we consider the logarithm in base 2:

$$Surprise = -log_2(p).$$

From this definition, we define uncertainty as the average surprise. Which makes intuitive sense: if we are surprised very often of what happens around us, it means we don't know much about the world we live in.

To understand the concept, we can take a very simple system –a coin.

In a coin, head and tail have the same probability. Let us imagine that for some reason we believe the coin to be biased. We believe that heads comes up 80% of the time and tails 20%. This means, each time head comes up our surprise will be

$$surprise(heads) = - log_2(.8) = 0.32 \ bit$$

and each time we see tail:

$$surprise(tail) = - log_2(.2) = 2.32 \ bit$$

Because the coin is actually not biased, we will have a surprise of 0.32 bit 50% of the times, and of 2.32 bit the remaining 50% of the times. On average, our surprise will be

$$Average\_surprise(we \ believe \ biassed \ coin) =$$
$$0.5 • 0.32 \ bit + 0.5 • 2.32 \ bit = 1.32 \ bit.$$

If we had believed that the coin was fair, as it was, our surprise for both head and tail would have been *surprise(head or tail) = − log₂(0.5) bit = 1 bit*

*average surprise would have been lower:*

*average_surprise(we believe fair coin) = 0.5 • 1 bit +0.5 • 1 bit = 1 bit*

*This will always be true: the average surprise is minimum when the probability we assign to each event is actually the frequency with which the event will happen.*

*More formally, we can say that if the system can be in N possible states, with an associated probability of $p_i$ and a frequency of $q_i$, our uncertainty S for the system is*

$$S = \sum_{i=1}^{N} -q_i \cdot \log_2 \left( p_i \right)$$

*According to the Gibbs' inequality, S has its minimum for $p_i = q_i$, i.e. when the probabilities we associate to each event, p, is the one we will actually observe, q.*

*In this sense, acquiring information on a system means knowing how to predict the frequency of each result.*

If we have a good model describing the solar system, we'll be able to predict the next eclipse and not be surprised when the sun will disappear.

A lion – a carnivore who is one of the laziest hunters in the animal kingdom – like any other living system works on minimising uncertainty, to get more energy (food) from hunting. Of various paths used by prey to get to a water hole, the lion studies which are the most probable (Schaller 2009). The lion minimise the uncertainty on the prays' path, and therefore increases the probability of extracting energy from the environment.

Note that there is no such a thing as absolute information. While the frequency with which events happens are not observer-specific, the probability we associate to them are. As Rovelli (2015) writes: "the information relevant in physics is always the relative information between two systems" (see also Bennett 1985).

## Information Storage Requires Energy

If the pages of this book looked something like this

few people would believe there was any message at all –rightfully, because this is just a randomly generated image.[4] It looks like the ink was just spread, randomly, into this state and then dried.

Viceversa, the Jiahu symbols, drawn in China 8600 years ago, are widely seen as one of the first example of human writing (Li et al. 2003).

Our intuition tells us that someone made an effort to create signs with a meaning, and that this drawing is not random. We don't know how valuable the information was, but we do know that someone wanted to communicate something, to pass information.

---

[4] https://upload.wikimedia.org/wikipedia/commons/f/f6/White-noise-mv255-240x180.png

To store information we need to set the system in a low-probability macrostate.[5] If we use as a communication tool a box with 10 balls and 9 walls, it would not be wise to assign a meaning to the state where all balls are evenly distributed: the system would often decay, if not controlled, to this state. A recipient would see nothing more than the "normal" state of the system.

This means that **information also needs energy**. Because we use low-entropy states, we need some energy to avoid the decay to equilibrium of the memory.

Random Access Memory (RAM) in a computer, for example, requires continuous energy input in order not to lose the information stored. Our brain dies in a few minutes without energy, and so on.

~

The last two sections bring us closer to understanding how a living system might work. They tell us that in order to absorb energy, the living system must store information about the environment. And that to store information, the system needs energy.

Information and energy

## Storing Information

We took as examples of information-storing devices writing, computers' memory and the brain. Only the last is (part of) a living organism.

---

[5] The opposite is not necessarily true, and there are indeed drawings, which we will analyse later, which are not universally recognised as symbols. This, because sometimes a system can settle itself in an apparently not-random state, to which we can assign a meaning, although no information was put there. For example, Italy, one of the largest exporters of footwear in the world, has the shape of a perfect boot. Nonetheless it's hard to believe that someone in the universe created the peninsula with such a shape to advertise the Italian shoe industry.

Nonetheless, for us humans, information is linked to language. We use language to store information, and to exchange information. Is the way we store information in language so different from how we living beings store information? Not really.

If we are asked about the structure of a brain, the first thing which comes to mind is probably "a network". Language is no different, as it is a network (Wood 1970). When we express concepts, we connect words. The information is stored by identifying patterns inside the network of language.

We could think of the book as a path in the network of the 6837 unique words[6] used to write it:

"The" ⇒ "amazing" ⇒ "journey" ⇒ "of" ⇒ "reason" ⇒ "\new_line" ⇒ "from"… and so on.

One characteristic of language is that part of the information is in the microstate – each word is different from the others– and part of it is in the macrostate –words have different meanings depending on which words they are connected to. Therefore part of the information is in the nodes of the network, part in the network itself.

What also is of interest, is that the complexity of the language network emerges as we use the language. The more we use language to describe new concepts, the more connections appear, the more complex the network becomes. One hundred years ago the word "so" was never used before the word "cool" and after "ur". Today, with usage, a new path has appeared.

Something similar happens in the brain. The memorisation process in the network of the brain occurs connecting nodes: "Pathways of connected neurons, not individual neurons, convey information. Interconnected neurons form anatomically and functionally distinct pathways" (Kandel 2013). As for language, information and complexity are inseparable in the brain.

Though the process is complicated (the human brain has various kinds of memory: short, medium, long-term, spatial, procedural), the memorisation mechanism involves the definition of new paths in the neural network, called Hebbian learning after the psychologist who came up with the theory in 1949.

Donald Hebb understood that memorising the association between two events (like in Pavlov's experiment, in which a dog associates the sound of a bell with food) occurs when two neurons fire several times simultaneously, so the activity of the same often appears to be related: the connection between these two neurons is reinforced, leaving an information trail (Kandel 2013).[7]

In the case of long-term memory, the link is definitive, in the case of short-term memory it isn't, and only remains if used frequently (by neurons that fire periodically). In general, with experience, the brain reinforces certain connections ("sensitisation through a heterosynaptic process", Kandel 2013) and weakens others ("synapses that are weakened by habituation through a homosynaptic process"). The less-used connections disappear, while the more-used ones are consolidated. Exactly in the same way as with language.

---

[6] Case-sensitive

[7] Specifically, "E3 – Mathematical Models of Hebbian Plasticity"

To sum up, networks are complex systems where their complexity, their internal structure, arise from storing information. In this sense, every network which stores information can be considered complex according to Herbert Simon (1962) definition of complexity: "Roughly, by a complex system I mean one made up of a large number of parts that interact in a nonsimple way. In such systems, the whole is more than the sum of the parts".

# Chapter 2
# From the Big Bang to Living Cells

We have seen that networks are complex systems, whose complexity, i.e. internal structure, arise from storing information. Living systems are information-storing devices, but so are other similar systems, like ant colonies or human societies.

Because we want to study the aggregation of matter from the Big Bang to the emergence of Artificial Intelligence, this chapter starts with a definition which encompasses all systems we will encounter in the book.

We then go on revising some theories on the emergence of life. Here we will see that molecular biology has identified communication as the engine behind the creation of the first biological cells.

## Intelligent Systems

Like a huge fractal, where each part is similar to the whole, life on earth appears to develop by subsequent degrees, creating aggregates of aggregates of said systems. It starts from proteins (aggregates of amino acids) through cells (aggregates of proteins) and complex organisms (aggregates of cells) to social networks – the superorganisms (Hölldobler and Wilson 2009) of ants and, lately, humans.

Internet isn't a living biological network, but it was created by biological organisms, exactly as biological cells created nervous systems. All these systems have the following properties which we observe from the outside:

- they're systems able to **store and process information**. The capacity of managing information is an emergent property –memories are systems in which information is recorded by connecting elements (proteins, cells, people, computer). The whole is more than the sum of the parts: therefore they are also *complex*, according to Simon's definition.

© The Author(s) 2020
M. Alemi, *The Amazing Journey of Reason*, SpringerBriefs in Computer
Science, https://doi.org/10.1007/978-3-030-25962-4_2

- they are **self-sustaining**: they can nourish their complexity autonomously. They have internal adjustment mechanisms, and manage to absorb sufficient energy from the external environment to maintain their internal structure and grow.
- they are **self-organising**: "A self-organising system is one in which a structure appears without specific intervention from the outside" (Griffith 2013). This concept derives from physics, but as the quote taken from a book on educational theory suggests, it's now commonly used.

There's no term in the theory of networks (see Appendix 1) to describe similar systems. We'll call them **intelligent systems**, with reference to the etymological origin of intelligence, from the Latin *inter-legere*: to join together, connect, in which *legere* has the proto-Indo-European root L-Gh that can also be found in the Italian word *legno* (wood), in the English word *log*, and in the Greek word λόγος (logos, word, speech, reasoning). The origin of the concept of *intelligence* is the ability to gather firewood, an essentially human activity used to extract energy from the environment: there are no known *Homo sapiens'* communities that don't use fire, and no other species of animals are known to use it. Being intelligent means *knowing* how to extract energy from the environment to remain self-sustaining and self-organising.

"Living networks" could also be a good definition: but not all intelligent systems can be considered biologically alive, although all can be traced back to biological life.

This race for ever-greater complexity, more information processed, more energy consumed, appears to be as relentless as it is necessary. Can we then understand why these systems emerge, and why they seem to have to evolve?

## Ex-Nihilo Energy and Information

Before we analyse the evolution of intelligent systems on earth, let's take a look at the creation of the universe – because all things considered, it's the structure of the universe that's responsible for the existence of our solar system and therefore also planet earth. Furthermore, the concept of information we considered in Chap. 1 is of late being seen as a possible key for interpreting the creation of the universe itself.

Every culture has its own creation myth. "Ours" (the scientific one) appears to be by far the most complex. But it has two big advantages.

First, it explains a lot of what we observe – not simply what we see when gazing at the stars on a dark night, but also what we see when we make protons collide close to light speed, and what we measure, with manic precision, with reference to the evolution of the universe.

For example: background radiation. In the universe, radiation appears to come from all directions. If you aim a satellite-TV dish tuned to 160 GHz in any direction, you'll receive an almost uniform signal. The two physicists who discovered this phenomenon by chance in 1964 won the Nobel prize, as they deserved to, because years earlier the existence of background radiation had been predicted as part of the Big Bang theory, in exactly the same way in which it was later measured.

There aren't many civilizations that aim satellite dishes at the night sky hoping to receive electromagnetic signals, and there are even fewer creation myths that can explain the existence of this radiation.

The second advantage of our creation myth is that it's based on very simple principles: interactions between matter are explained by so-called spontaneous symmetry breaking. As the temperature of the universe drops, asymmetry appears in the form of new forces.

Spontaneous symmetry breaking in quantum field theory isn't easy to understand, but the phenomenon that's usually taken as an example, is. Imagine a red-hot ball of ferromagnetic material. At high temperatures, the material won't be magnetised. As the temperature drops, the ball becomes a magnet, with the magnetic field in some random direction.

This means that by lowering the temperature we can *acquire information* on a system of balls. In a system consisting of many balls, all above critical magnetisation temperature, we cannot distinguish one ball from another. Below the critical temperature however, each ball will be different to the next: as the direction of the magnetic field of a ball is casual, so each ball can be identified by its magnetic direction. In practice, the difference between macrostate and microstate collapses: from maximum it becomes minimum, and the system "lets us" describe it in a better way, with more information, i.e. less uncertainty.[1]

Similar mechanisms lie behind the laws that have governed the universe since its primordial explosion, the Big Bang, 13.8 billion years ago. The most interesting thing is that in the last 50 years, in a more and more decisive way, temperature (in other words energy density) and information have been used to probe the "metaphysical": the very origins of matter-energy and time that constitute our universe (Tryon 1973), (Vilenkin 1982), (Wheeler 1990), (Lincoln 2013).

Information and energy, in these theories, are not linked merely on a thermodynamic level, but at a much more intimate level: energy represents the other side of the coin to information. Having information means having energy, and it's *the spontaneous creation of information that represents the basis for the emergence of matter in the origin of the universe.*

Although the creation of the universe remains within the scope of metaphysics, one of the merits of these speculations is interpreting energy in terms of information, and therefore probability: something that's even more fundamental than energy itself. Probability, in fact, seems independent of the type of universe in which we happen to live: we could imagine a universe in which the fundamental interactions are different, but it's difficult to imagine a universe in which the probability of coming up "heads" when tossing a fair coin isn't 0.5.

---

[1] As we've already seen, entropy should be considered a property of certain macroscopic variables (those that describe the macrostate), and information is always relevant information between two systems (Rovelli 2015)

## The Emergence of Complexity

In Chap. 1 we saw how complex systems become all the more complex as the amount of information they contain increases, but the first great complex system actually emerged spontaneously: is the universe, or cosmos, rightfully from the Greek word κόσμος (kosmos, order).

Around 300,000 years after the Big Bang, the universe was an immense, rapidly expanding cloud of hydrogen, with just a little helium and a few other elements. We can compare this gaseous mass to a random network (Erdös and Rényi 1960): each atom influences and is influenced by only a small, almost constant number of other atoms.

In practice, there is maximum entropy in the system, because all the elements – the hydrogen atoms – are indistinguishable inside the network (see Appendix 1 for how to compute the entropy of a network). It seems like the end of the universe: as every system slips towards greater stages of entropy, and as the universe is now at a stage of maximum entropy, it should remain as it is and not evolve.

In reality, a system cannot become ordered unless it is supplied with energy. And although the universe cannot receive external energy, it is a system in unstable equilibrium. The entire universe cannot "recollapse" in a Big Crunch because of the explosive power of the initial Big Bang, but areas in which the gas becomes more compact can form on a local scale.

Imagine the explosion of a bomb made up of magnetic parts: even if the average distance of all the parts from the centre of the explosion continues to increase, some parts will still come together. In practice, instead of many small parts moving away from the centre of the explosion, there will be fewer, bigger ones.

Something similar occurs in the universe. The gas cloud collapses in various points. High-density gas aggregates under extremely high pressure and forms the stars. Stars are indeed compressed hydrogen that fuses into heavier elements in the cores. The fusion produces heat, which stops the stars from collapsing.

The first effect of the formation of stars is the creation of thermal gradients – areas at high temperatures (the stars) and low temperature (everything else). One can extract energy from such a system with any thermodynamic engine, i.e. something which extract heat from the hot reservoir and releases it to the cold one, transforming part of it in mechanical engine.[2]

The second effect is the **spontaneous emergence of complexity**. Not all stars are created equal: some are very rich in mass, heavy (few) and others are very poor, light (many). In other words the size of stars has the same structure of the wealth of people in a society does, the so-called Pareto distribution (or power-law, see Appendix 1). Furthermore, stars form aggregates, galaxies, in which the number of stars follows the same distribution (few really big, many very small), and galaxies form aggregates of galaxies (clusters) with the same Pareto distribution.

---

[2] Steam and internal combustion engines work in this way, see for instance https://en.wikipedia.org/wiki/Heat_engine.

In short, something incredible is happening: matter organises itself and the structures typically found in intelligent systems appear: Pareto distribution and self-similarity.

The emergence of the Pareto distribution in this case has nothing to do with the process of storing information – it's a "cosmic coincidence" (Watson et al. 2011). When a glass breaks on the ground, the size of the pieces also follow the Pareto distribution, but there's no form of intelligence behind the process (see Appendix 1 for why this happens).

Intelligent process or not, the universe has now a structure which makes possible the creation of more complex matter, beyond simple hydrogen and helium. In big stars, gas is subject to greater pressure, and burns faster. The few macro-stars burn all their fuel in a few billion years, and explode. Just before the final explosion they produce all the elements in the periodic table. Supernovas, as they are called, have been throwing elements out into space for 12 billion years. One supernova is also at the origin of the matter from which the planets in our solar system are made, and therefore us too.

In physics, the model that describes the formation of all elements – from hydrogen to uranium – during the life of the universe, was developed in the 1950s, and called $B^2FH$, after the authors' initials (B2FH 1957).

The prodigiousness of the $B^2FH$ model is that it not only explains what we observe in the universe. It provides us with an image of a universe in continuous change also in chemical terms. New elements continue to be formed with the passing of time, and the abundance or scarcity of elements lets astrophysicists calculate the age of various regions in space. It is estimated for instance that our solar system, earth included, formed 4.56 billion years ago.

That's when life could emerge.

## Life Without Selection

Charles Darwin stated that asking questions about the origin of life was the same as asking questions about the origin of matter. This did not stop him imagining scenarios which, he thought, might have favoured the formation of living organisms: "some warm little pond with all sort of ammonia and phosphoric salts" Peretó et al. 2009).

Darwin was even criticised by his admirers for not having proposed an official theory on the origin of life (Peretó et al. 2009). We must give him credit for having avoided taking up the challenge. In the *On the Origin of Species*, Darwin described the evolution of a few complex organisms, made up of biological cells, the structure of which was unknown at the time. It was before Gregor Mendel discovered the laws of inheritance (1865), and of course before DNA was even imagined.

Darwin probably didn't propose a theory for the origin of life simply because applying Darwin's mechanism of natural selection to the emergence of life, as done by Dawkins (1976), is like comparing apples with pears (Johnson 2010). What's more, the idea that a self-replicating molecule with an information content *casually* appeared in a primordial soup, as imagined by Dawkins (1976) ("At some point a particularly remarkable molecule was formed *by accident*. We will call it the Replicator.") appears to be statistically groundless (Yockey 1977).[3]

The fact that the "Replicator" cannot have appeared merely by chance has been considered proof that there must be an intelligent design behind it. As usual, whenever there is no clearly valid scientific explanation, the *intelligent designer* comes in. When a plausible explanation becomes common sense, as in the evolution of the universe, the *intelligent designer* retreats.

Consider a statement like the one John C. Eccles made just in 1989: "Since *materialist solutions fail to account for our experienced uniqueness*, I am constrained to attribute the uniqueness of the Self or Soul to a supernatural spiritual creation."(Eccles 1989). Today, scientists think that purely materialistic models can provide a perfect explanation for the emergence of conscience (Dehaene 2014), and few people, let alone a scientist, would defend Eccles' idea.

In a similar way, our knowledge of molecular biology today leads us to consider the creation of life through divine intervention, or the emergence of the same "by accident" as in (Dawkins 1976), not to be the best choice.

The question is therefore: with the instruments at our disposal, can we describe the formation, evolution and behaviour of various intelligent systems, from biological cells up to human societies, going through the nervous systems?

## The Startups of Life

There is often some confusion about physics dictating that "it's impossible to create order" or "disorder is constantly increasing." We have seen the universe itself created order soon after the Big Bang. When we freeze water in a freezer we create order. It's just that we need energy to do so.

Evolving systems can be divided into 4 categories:

1. The system becomes more ordered and absorbs energy – the "freezer" system. Possible only if we provide sufficient energy.
2. The system becomes more disordered and emits energy – the "explosion" system. Possible only if there is not too much energy emitted.
3. The system absorbs energy and becomes disordered – the "adolescent" system. These reactions are always possible. It's easy to waste energy for creating disorder!
4. The system becomes more ordered and emits energy –the "genie in a bottle" system. Unfortunately, it's impossible.

---

[3] There is still the possibility that life appeared on earth after this casual process had occurred an infinite number of times: both in an infinite number of universes and in this universe, and billions and billions of times in systems similar to our solar system. It's possible, but perhaps it would be better to come up with a mechanism that makes us less unique...

Figure 2.1 shows how the systems above evolve. They all start from a state where no energy is absorbed or emitted, and the order is not perturbed. They then start evolving, and finally stop. The fridge is switched on, absorbs increasingly more energy, creating more order, then stays in a region where it absorbs an almost fixed amount of energy, creating a fixed amount of order. Until it's switched off.[4]

But if we take life since the very beginning, as shown in Fig. 2.2, there are two things which do not fit.

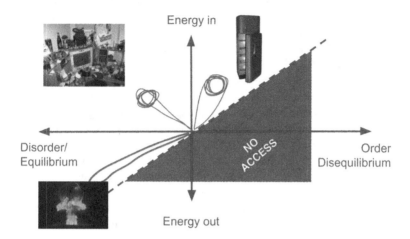

**Fig. 2.1** The three possible evolution of a system in the space "energy absorbed/emitted" versus "order created/destroyed"

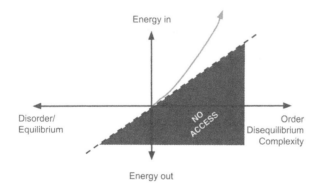

**Fig. 2.2** Since its appearance, life has been absorbing increasingly more energy from the environment, constantly escaping equilibrium, i.e. creating more order

---

[4] Pictures available under creative common licence: https://www.needpix.com/photo/download/986345/refrigerator-freezer-fridge-freezer-retro-seventies-american-style-metalic-cool-cold, https://www.flickr.com/photos/puyo/253932597, https://www.flickr.com/photos/51686021@N07/33230171512

The first is –why it started? A fridge starts cooling, when we provide energy, because that's the way we built it. We are the intelligent designer.

The second is –why it keeps growing? "Because life evolves" is not a good answer. "Because of natural selection" isn't either: why should natural selection bring an increase in energy absorbed and order?

To answer, let's have a look at the ecosystem of companies. Specifically, startups growing without external funding (the ones who can "bootstrap",[5] a concept we will encounter again later). A startup which bootstraps has little or no internal organization, just a few elements, but is able to absorb some energy (cash from sales).

If the business model, i.e. the information retained by the founders on how to extract energy from the market, is valid, the startup will use some of the earned cash to increase in size. Together with that, internal organisation –complexity– will appear.[6] The organisation represents an asset for the company –is what allows the exploitation of the market– but also a liability –it requires energy in the form of cash.

The startup needs little cash at the beginning while providing, with its "disruptive idea", a way to extract revenues from the environment. The extra cash will be used to improve the internal organisation, while increasing revenues. It's an avalanche effect: more organisation increases the revenues, bringing more organisation, which requires more cash but allows more revenues and so on.

With growth, there is a scale effect: bigger networks –companies– can process more information per element, notwithstanding the fact that the information processed by each element is less than it was at the beginning. It's Herbert Simon definition of complexity –the whole is bigger than the sum of its parts.

In a similar way, we must find a mechanism which allows the aggregation of basic chemical elements into something more complex through absorption of energy. The information on how to extract energy must be contained somewhere, and these elements must be able to "read" it. Once the elements get together, they must be able to store information on how to aggregate further, storing increasingly more information.

It looks like an impossible task but, seen that we are here, alive, it must be not.


## Amino Acids – The Entrepreneurs of Life

Amino acids are big molecules which, bonded together, form proteins, which eventually organise themselves into biological cells. One of the most interesting discoveries in biology during the last century was that amino acids, considered the building blocks of life, appear spontaneously in nature.

---

[5] "Building a company from the ground up with nothing but personal savings and, with luck, the cash coming in from the first sales". https://www.investopedia.com/terms/b/bootstrap.asp verified 26 August 2019.

[6] Lex Donaldson, a sociologist, defines complexity as "a measure of the amount of knowledge available within the organization (Donaldson 2001)

Soon after World War II, Harold C. Urey, a Nobel prize and key figure in the development of the nuclear bomb, began to consider the emergence of life in an atmosphere rich in methane, hydrogen, and ammonia (Bada and Lazcano 2003). It's Darwin's "primordial soup".

Urey's idea had such an impact on the 22 year old Stanley Miller, another physicist, student of Edward Teller – often referred to as the Father of the hydrogen bomb[7] – that he decided to abandon nuclear physics for setting up an experiment which proved Urey's hypothesis.

In 1952, for the first time, humanity produced synthetic amino acids in the Miller-Urey experiment.

Today, we believe primordial conditions were different to those imagined by Urey and Miller, but the fact remains that the formation of amino acids can be a spontaneous process. It's a natural "freezer system", a little order is created and in exchange a little energy is absorbed.

But how amino acids, these big molecules, can aggregate in more complex structures, like proteins? This cannot be a spontaneous process. To bond amino acids we need much more than just electrical discharges: they require specific chemical processes. And these chemical processes do not occur by chance, especially on the necessary scale. In addition to that, even if we managed to supply energy in the right way, we couldn't create long amino acid chains – proteins – as we wished, because they would break.

One mechanism that partially solves the first problem was proposed soon after Miller's experiment (Koshland 1958). There are substances, called enzymes (from the Greek word ἐν-ζυμον, in yeast) which, due to their particular form, and therefore the electric field around them, can favour the creation of a bond between two amino acids, the peptide bond.

It's like building a chain of lockers with or without their keys – energy is required in both cases, but much less in the first case. In presence of enzymes, amino acids bond easily, using just some excess of the energy available in the environment.

There is still the second problem: it's impossible to have proteins, chains of billions of amino acids, free in the environment. Imagine a human chromosome – a DNA molecule – lying unfolded: it would be two metres long. In any natural environment it would break immediately.

This problem is solved by collaborative molecules, called "chaperones". Molecular chaperones can favour, without having to do any work, the folding of the proteins into stable configurations. These stable configurations have the additional property of favouring proteins interaction. When folded, proteins can move in water, interacting with each other without getting trapped.

Egg white is an excellent example of how important protein folding is. When raw, it's fluid, and consists of an aqueous solution of folded proteins. When beaten into a thick foam or boiled, the white solidifies because the proteins burst and are

---

[7] For his obsession in building such a destructive device, together with his strong Hungarian accent, Edward Teller is also the person the character Dr. Strangelove in Stanley Kubrick's film was based on.

trapped in each other. In the first case the proteins can move around and interact, in the second they are immobilised.

In conclusion, for proteins to form, survive and eventually interact we must have:

1. A mechanism which creates enzyme for amino acids to bond together in proteins
2. A mechanism which creates chaperones, for the proteins to fold

## The Secret of Life

Ever since it was discovered, DNA has been called "the molecule of life". But really, it would be more fitting to call it "the brain of cellular life". Our brain, while essential for our survival, could not survive alone. The same is true for a DNA molecule, which alone would remain helpless and in time disappear.

Nevertheless, the idea that DNA is key not only for cellular life, but also for the appearance of life itself, has never been far from the thoughts of many chemists and biologists alike: one of the most popular theories on the origin of life is in fact the so-called "RNA world", in which ribonucleic acid, a molecule similar to DNA, was responsible for having kick-started life.

The "RNA world" hypothesis derives from the fact that, in a similar way to Miller's experiments, we've managed to reproduce the conditions in which amino acids organise themselves and bond in RNA chains which, in the right conditions, start reproducing. The only problem with this theory is that in any case, at the start of the process, you have to introduce enzymes of biological origin. Moreover, DNA and RNA are used by cells to store and transport information on how to construct proteins, so we don't know how an RNA that reproduces autonomously could evolve into cells (Robertson 2010), (Davies 2000).[8]

Another hypothesis for the origin of life, better-suited to the framework of intelligent systems, is the "proteins first", proposed by physicist Freeman Dyson (2004). Going against the idea that life started by creating both the information medium and the information itself (RNA) *at the same time*, Dyson took inspiration from the theory of Russian biologist Alexander Oparin (1957). Oprain had the intuition that life is made possible by how molecules interact. Life begins with the interaction between proteins, and not with their capacity to reproduce.

What's fascinating about Oparin's view, is that life is seen as a natural process in the evolution of matter. Previously we considered how the universe created more and more complex structures, starting from a shapeless *blob* of energy, the Big Bang. In Oparin's opinion, life should be considered the continuation of this process – a continuation that stopped at very low levels in some parts of the universe, while on earth and perhaps on other planets it continued to create complex organisms.

---

[8] "Is Eigen's work a reconstruction of the steps in which nature created life from inanimate matter? Evidently not. "Page 139 in the Italian version.

The idea that life started thanks to the fact that amino acids spontaneously became organised into complex structures has become all the more credible in recent years thanks to theoretical and experimental works. Guseva et al. (2017), for example, introduce a complexification mechanism based on the polarity of water molecules and polymers.[9]

The mechanism is ingenious, and surprisingly simple: the authors prove that in certain conditions a small percentage of proteins[10] fold in water, because some amino acids are hydrophilic and others are hydrophobic: the first tend to form the surface of the folded protein, while the second form the internal part. Some hydrophobic amino acids remain exposed though, attracting the hydrophobic elements of other proteins.

Sometimes two hydrophobic tails fall together into the field of another hydrophobic section on the surface of the refolded protein. Like a magnet that attracts other magnets so they stick together, this section helps the two tails interact and form a bigger protein – it's the enzymatic process. In practice, small proteins become the enzymes that help other proteins grow ever bigger. The same mechanism helps these new proteins fold.

Some proteins, highly interacting with each other, combine spontaneously in closed communities, called protocells, that can attract or search for the substances they need to function. The fact that a few molecules combined, *protocells*, can form systems able not only to use the chemical energy available in their environment, but also to move in order to find it, has been proven in various experiments (e.g. see Hanczyc et al. (2003), made famous by Hanczyc's TED talk[11]).

Protocells are dissipative structures: structures that can use an energy flow to create and maintain an internal order (Kondepudi and Prigogine 1998), (Margulis and Sagan 1997). When there is no energy, the internal structure disappears, exactly as an organism without food decomposes.

While remaining within the field of speculation – there is still no widely accepted paradigm for the origin of life – according to Dyson at this point the RNA can form (as there are enzymes) and reproduce (the RNA can produce its own reproductive enzyme, the ribozyme).

The RNA initially appeared as a parasite. After having exploited the spontaneous emergence of simple polymers/enzymes, it started to grow, absorbing the same polymers it no longer needed thanks to the ribozyme, destroying the host protocell.

At this point, the RNA had two options: continue to emerge spontaneously, grow, and destroy every protocell it manages to infect, and therefore disappear, or collaborate with the protocell on which it is the resident. Since 1982, in fact, we know that

---

[9] A term used generically for chains of simpler molecules, the monomers. It includes also, but not only, proteins.

[10] Specifically, we are talking about very small proteins, so small that they are called peptic chains and not proteins.

[11] Hanczyc, M. M. (2011) The line between life and not-life. TEDSalon London Spring 2011. https://www.ted.com/talks/martin_hanczyc_the_line_between_life_and_not_life

RNA acts in cells not only as an information medium, but also as a protein enzyme: in other words it helps other proteins form.

In the words of Freeman Dyson:

> Within the cell, with some help from pre-existing enzymes, the nucleotides produced an RNA molecule, which then continued to replicate itself. In this way RNA first appeared as a parasitic disease within the cell. The first cells in which the RNA disease occurred probably became sick and died. But then, according to the Margulis scheme [which we'll discuss later], some of the infected cells learned how to survive the infection. The protein-based life learned to tolerate the RNA-based life. The parasite became a symbiont. And then, very slowly over millions of years, the protein-based life learned to make use of the capacity for exact replication that the chemical structure of RNA provided. The primal symbiosis of protein-based life and parasitic RNA grew gradually into a harmonious unity, the modern genetic apparatus. (Dyson 2004)

Dyson adds that what his is "not yet a serious scientific theory". One might add, nonetheless, that the idea that life followed what happens at a macroscopic level – first the medium for storing the information (hardware) is supplied, and then the information itself (software) – appears to be more plausible than the RNA world theory.

The idea that a parasite can turn itself into an information processing centre is incredibly appealing and, far from being a phenomenon specific to cells evolution, it's something we come across in nature –and in our society, as we will see– all the time.

A parasite is an organised structure that grows in another organised structure, in a way that's detrimental to the same. The RNA parasite found its way into cells – fragile structures always looking for new mechanisms to supply the necessary energy – finally destroying both itself and the cell.

It would be much better, for the protocells, to learn how to use this fantastic molecule, and use it on the one hand as an enzyme, and on the other as an information medium to memorise which proteins were necessary, and when.

With the inclusion of RNA also came the ability to replicate, and therefore also evolution through natural selection. Selection, for the reasons explained above, cannot lay the foundations of life; it's just one of the processes that allows life to evolve. As the physicist Jeremy England said, "you might call Darwinian evolution a special case of a more general phenomenon" (Wolchover 2014).

The secret of life is collaboration, not selection.


## Evolution Through Learning

Even if we accept that proteins managed to absorb an RNA molecule and use it as a data processing centre, one fundamental question remains unanswered: how did protocells with RNA evolve into cells able to duplicate their "brain" and pass it down to descendants?

So we must imagine a mechanism that drives life towards a stage of *evolvability,* or a propensity for evolution (Watson and Szathmáry 2016).[12] For example, the

---

[12] See "Outstanding Question 1"

ability to reproduce is essential for the evolution of organisms. By reproducing, an organism increases its *evolvability* also thanks to natural selection, and its code will be favoured over others (reproduction also favours evolution because, as we'll see below, makes it also possible to reset unnecessary complexity).

We saw how Hebbian learning works in the brain (Chap. 1, *Storing information*): two neurons that are activated together reinforce their connection. The same thing is true for language, and it's a classic example of unsupervised learning in Artificial Intelligence: if the words "kitten" and "cat" are often used in the same context, a computer (or a person) understands they are synonymous – i.e. they are connected – without needing external training.

But we've also seen that this mechanism is the one many complex networks use for memorisation. Not only neural networks, but also societies (two people who have the same interests are likely to form a bond) store information as Donald Hebb described in the 1940s. Creating paths between the elements, and assigning a meaning to each path.

The same thing probably happens with genes. Exactly like the brain doesn't memorise in terms of single neurons (the "words"), but rather in terms of the connections between the neurons (the "phrases"), a cell's information is stored in a genetic network, the GRN (Genetic Regulatory Network) (Davidson et al. 2002)

The cell learning mechanism is essential for adaptation in unicellular organisms, and is used for the formation of complex organisms. How can two cells with exactly the same DNA become part of such different parts of the body as the intestine and the brain? By "learning" what they must become. Learning doesn't necessarily have to be Hebbian, as there are non-Hebbian mechanisms that let the cell inhibit or favour certain genes (Epigenetics 2013), but Hebbian learning was recently shown to be a possible learning mechanism (Watson et al. 2010), (Watson and Szathmáry 2016).

## Artificial Neural Networks and DNA

The Hebbian model was also the model that inspired the first attempts to replicate the learning capacities of biological neural networks with computers, although without much success (Anderson and Hinton 1981).

The situation changed in 1982, when physicist John Hopfield proved that it was possible to define an "energy" function for mathematical structures resembling a biological neural network.

Physical systems tend to settle into low energy states, like a ball in a bowl. If we have a function describing the energy of a system, we can also identify stable configurations. This is what Hopfield did: he proposed a function computing, for any configuration, a certain variable, which he called "energy". Starting from an input, it was possible to find, always, a particular configuration minimising this "energy" function. Similar inputs would have the same minimum, and therefore would correspond to the same "stored message". It's like saying we can recognise similar bowls because the ball will move to the same position however we throw it in (Hopfield 1982).

In practice, Hopfield networks are defined by the weights of the relationships between the neurons. The energy of a neuron is the sum of the weights for its state of excitation (0 or 1). The energy of the network is the sum of the energies of each neuron. When each neuron tries to minimise its energy value, independently of the others, a Hopfield network falls into a configuration (neurons activated / deactivated) that minimises the energy of the whole network.

To store a vector [a list of 0 and 1s], the weights are changed to reduce the energy of that vector. So the stored vector correspond to the local minima in the 'energy landscape'. To retrieve a vector from a noisy version of it, the network is put into an initial state that represents the noisy version and then allowed to settle to an energy minimum (Hinton 2014).

But what have Hopfield networks – a computational neuroscience instrument – got to do with the genetic code? A lot! The genetic code shouldn't be seen as a mere *storage device*, dumb memory. It's more similar to our brain than to a hard disk. Paul Werbos, the mathematician who in his 1974 graduate thesis showed how we can "make artificial neural networks learn", wrote:

...to what extent is the "junk DNA," 97% of the genome, actually a kind of learning system, like the brain. More precisely, how much of the genome is intended to help us learn how to choose better gene expressions, as opposed to merely specifying the final "actions" at the output layer of the "gene brain"?[13]

Geoffrey Hinton, probably the most important figure in the development of artificial neural networks, as long ago as 1987 wrote, together with Steven Nowlan:

Many organisms learn useful adaptations during their lifetime ... It seems very wasteful not to make use of the exploration performed by the phenotype to facilitate the evolutionary search for good genotypes. The obvious way ... is to transfer information about the acquired characteristics back to the genotype (Hinton 1987).

While today epigenetics have legitimised the idea that genetic code can pass information acquired during the life of the organism on from parent to offspring, this wasn't so obvious in the 80s, and John Maynard Smith – one of the most important geneticists and scholars of evolution of the century – had to publicly support Hinton's (1987) publication when this was blocked from publishing (Smith 1987). The concept, taken up again by Watson and Szathmáry (2016), is clear: to evolve, a successful system must be able to learn, and pass what it learns on to its offspring.

A living system which, when it dies, loses all the information it acquired during its lifetime would not only pointlessly destroy something precious, but also slow the evolution of the species to such an extent that it would soon become extinct (Hinton 1987).

According to biologist Lynn Margulis (1997), to whom next section is dedicated, a crucial feature of living systems is that the internal organisation they support is simply the information necessary for finding the energy they need –exactly what we discussed in the previous chapter.

---

[13] https://www.werbos.com/life.htm, retrieved August 10, 2018.

In a more general framework the ultimate purpose of every intelligent system, including living systems, is to reduce uncertainty concerning the external environment, storing information, to acquire the energy necessary to sustain its internal order. It's therefore only natural that life has adopted a data processing centre, DNA, that lets it learn, and therefore quickly evolve.

## Collaboration and Eukaryotes

The Dyson quotation in the previous section refers to the "Margulis scheme". Dyson explains that he got the idea from a theory created to explain the emergence of more complex biological cells, called eukaryotes, published by Lynn Margulis (1970).

Dyson (2004) also wrote a short, excellent description of Margulis's original theory:

> ...the evolutionary tree has three main branches representing a divergence of cell types far more ancient than the later division of creatures into animals and plants. Moreover, the genetic apparatus carried by organelles such as chloroplasts and mitochondria within eucaryotic cells does not belong to the same main branch of the tree as the genetic apparatus in the nuclei of eukaryotic cells. The difference in genetic apparatus between organelles and nucleus is the strongest evidence confirming Lynn Margulis's theory that the organelles of the modern eucaryotic cell were originally independent free-living cells and only later became parasites of the eucaryotic host.

> According to this theory, the evolutionary success of the eucaryotic cell was due to its policy of free immigration. Like the United States of America in the nineteenth century, the eucaryotic cell gave shelter to the poor and homeless, and exploited their talents for its own purposes. Needless to say, both in the United States and in the eucaryotic cell, once the old immigrants are comfortably settled and their place in society is established, they do their best to shut the door to any prospective new immigrants.

With his reference to politics, Dyson suggested that the Margulis scheme – the idea that parasitism and symbiosis are the driving forces behind evolution – could help understand not only the emergence of life but also properties of human societies which should be analysed in terms of emergence properties.

## The Importance of Scientific Revolutions

When Lynn Margulis died in 2011. The obituary published by Nature praised "her paradigm-changing book, *Origin of Eukaryotic Cells*", published in 1970 (Lake 2011).

The term *paradigm* refers to Thomas Kuhn's (1962) *The structure of scientific revolutions*. In his work, Kuhn, a physicist and historian of science, proposes a social vision of the scientific method. There are no strict rules on what is scientific and what is not (contrary to what Karl Popper (1934) thought). It's a *shared idea,* which Kuhn calls *normal science*, that guides a scientist's work. There are works, in this normal science, that can be considered paradigms (from παρα-δείκνυμι, show to the side). Aristotle's *Physics* in this sense is no less a scientific work than Galileo's *The Assayer*, Newton's *Principia*, or Freud's *The Interpretation of Dreams*. All of these works coagulated the knowledge of the time in order to refer to a multitude of concepts, connected with each other, without having to explain what the author is referring to all the time. The concept of force in physics was crystallised by Newton, that of the unconscious by Freud. Every time we use these words, we're referring to the Newtonian and Freudian *paradigm*.

As Feyerabend (1989) then noted, a shared paradigm can eventually stiff the evolution of scientific thought. There will be scientists who'll defend it even if it's indefensible: Aristotelians, like the character of Simplicio in Galileo's *Dialogue*, are the problem, not Aristotle.

The anatomist who in the *Dialogue* observes that all nerves originate in the brain but remains convinced they originate in the heart because Aristotle said so, strikes a sour note ("You've shown me this thing which is so open and sensible, that if it weren't for Aristotle's text of the opposite opinion, as it openly says, the nerves originate in the heart, one would have to confess it was true").

But considering what we have discussed about uncertainty and coins in Chap. 1, the anatomist behaviour might be justified. As we've seen, considering a coin as fair, i.e. that it will have a 50% probability[14] of coming up heads or tails, is the safest choice we can make. We have held thousands of coins in our hands, and none of them seemed biased. If, without an absolutely valid reason, you were to consider the coin biased with an 80% probability of coming up heads and a 20% probability of coming up tails, you'd risk increasing your uncertainty and if taking bets you'd lose money fast: the bet placers would continue to consider the coin fair, and rightly so, while you would take ridiculous bets (such as "heads" 3–1) and lose time and time again.

On the other hand, if the coin is actually biased but everyone, including you, thinks it's fair, no one would win much as everyone has the same average uncertainty. Likewise, if the consensus for 2000 years was that nerves originate in the heart, stating that the opposite is true means risking everything. Just before *Dialogue* was published, Jacopo Berengario of Carpi, the first ever neurosurgeon, was called to operate on the fractured skull of Lorenzo de' Medici, one of the most powerful men in Renaissance Italy. If Berengario had used Aristotelian medicine and Lorenzo had died, his reputation would have suffered. If Lorenzo had died while being operated on by a non-Aristotelian surgeon, as Berengario was, he would have lost not only his reputation but also his head.

The problem was that if he had used the Aristotelian nerve model, the whole society would have lost. The "anti-information" on the origin of nerves caused the deaths of many patients, and it was down to scientists like Galileo and the likes to

---

[14] We always refer to the Bayesian definition of probability: "Probability ... serves to express, in a precise fashion, for each individual, their choice in their given state of ignorance" (de Finetti 1970)

put forward counterarguments convincing enough to destroy the old paradigm and create a new one.

Society will benefit from a new paradigm that minimises uncertainty, and can create new instruments to sustain its internal complexity. It's the same mechanism Margulis (1997) described for the evolution of cells, and can be applied to her own work.

Tenaciously sticking to the paradigm can be counterproductive. For example, using the principle of natural selection as the driving force of evolution of societies – as Darwin himself and many others did – leads to make hazardous conclusions. According to the English naturalist, hospitals and health services were the cause of the decline of our species: "We ... do our utmost to check the process of elimination; we build asylums for the imbecile, the maimed, and the sick; we institute poor-laws ... Thus the weak members of civilised societies propagate their kind ... *this must be highly injurious to the race of man. It is surprising how soon a want of care, or care wrongly directed, leads to the degeneration of a domestic race; but excepting in the case of man himself, hardly any one is so ignorant as to allow his worst animals to breed."* (Darwin 1874).

Darwin was wrong. There was a period, after the end of the Second World War, in which the human race and its living conditions improved, not thanks to selection but instead thanks to social policies and collaboration (Pinker 2011), (Norberg 2016). Fascists and Nazis, who followed social darwinism and wanted to suppress what they considered the "weaker" elements of society, produced as a result hellish conditions for human beings.[15]

Margulis proved that life emerges as a result of collaborative processes: collaboration is the driving force behind evolution.[16] This is what the present book wants to take to extremes: collaboration – in practice the exchange of information/energy – lies behind every successful long-term enterprise, from eukaryotes to nation-states. And that's not all: the union of various elements is what makes it possible for everything to store information. If humanity was composed of only the "fittest" it would fail, as would a language made up of only particularly eloquent words. Obviously, every new word can only enrich the language.

What's more, as mentioned in (Margulis 2008), Darwin's *On the Origin of Species* isn't about the origin of life or even the origin of the species.[17] Not acknowledging that natural selection plays a role in the origin of life doesn't make it any less important for the evolution of the species, but "pure" Darwinists seem unable to accept other mechanisms in biology, sociology and whenever they believe selection

---

[15] With the help of a Darwinian foreign policy aimed at defending living space, *lebensraum* (Ratzel 1901), rather than collaborating with other states.

[16] The fact that Margulis was worthy of an obituary in Nature, which called her book paradigmatic, therefore measuring it with the same yardstick as Darwin's work, did not stop her book from failing to be acknowledged by anyone outside a close circle of scientists, and, unfortunately, being literally eradicated from human memory. No further editions of the book were ever printed and finding an electronic version is impossible.

[17] As the subtitle of Darwin explains: it's about "the Preservation of Favoured Races"

could be applied. Let's not forget that Darwin broke ties with the past by valiantly defending the concept of evolution: it's not natural selection that's changed the way we think about life on earth, but rather the idea that life evolved.

If this change occurred thanks also to collaboration and not just selection, this means that a collaborative societies, whatever Darwin thought, might not only be nicer to live in, but also more likely to be successful.

# Chapter 3
# From Complex Organisms to Societies

We have seen how the evolution of the universe brought exploitable energy, complexity and finally, at least on our planet, life. In this chapter we will see why life had to evolve beyond unicellular organism, creating complex organisms and then aggregates of those organisms –societies.

## Intelligence Needs Energy

In order to understand the evolution of life in terms of the subjects covered in Chap. 1, let's take another look at the example of the container with a semi-divider wall between two compartments:

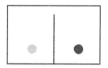

If the system is microscopic, thermal disturbances will make balls pass from one side to the other. We can make the system more stable by using heavier balls, but in this case we'll need more energy to change the state of the system and store some information.

The system with light balls is more efficient, but also fragile. The other is stronger, but inefficient. Another trick could be using several systems of light balls to store the message in a redundant way, but in the end we'd have to use just as much energy as in the heavy ball system.

In nature there are robust "heavy ball" information tools such as DNA and other more agile "light ball" ones like the nervous system. All of them need energy to be set or remain in a particular state. This is not related to the information the state carries.

© The Author(s) 2020
M. Alemi, *The Amazing Journey of Reason*, SpringerBriefs in Computer
Science, https://doi.org/10.1007/978-3-030-25962-4_3

We call a particular state to which we associate a meaning as *syntax*. Syntax has to do with the structure of the message, it is the substrate on which we put a message, but has nothing to do with its value. With 5 bits we can pass on a message that reduces uncertainty (the next number on a roulette wheel where the ball will land, see Appendix 2) or not (a number where the ball won't land). But to store these 5 bits we need energy.

As it costs energy to build and use syntax, regardless of its utility, every intelligent system has to make sure every processed bit brings information, i.e. reduction of uncertainty about the environment. Intelligent systems must be able to change their mind according to observation.

We do that quite easily for simple cases. For example, how likely is it that a coin is biased if after 100 tosses it comes up heads 51 times and tails 49? Not very likely, as there aren't that many biased coins around and the ratio of heads to tails we observed is close to 1. But if it comes up heads 89 times and tails just 11 it makes sense to consider the hypothesis that the coin is biased as likely.

This seems obvious, but while it's relatively easy for one brain to change opinion, it's not so easy for a network of brains to do the same. We quite rightly refer to a company's *DNA*, not its frontal lobes. "Organizational capabilities are difficult to create and costly to adjust" (Henderson and Clark 1990), in the same way as for DNA.

Unfortunately, an organisation created to extract energy needs itself energy. Not updating the knowledge – keeping the same syntax even when it's no longer of value, when it produces no more energy – means accepting to keep an energy investment portfolio that will result in a loss.

This is also what ageing means: storing syntax that provide less and less up-to-date information. Evolving, on the other hand, means updating the stored information on the basis of the changes in the environment we are living in. Life and evolution are a continuous compromise: attempting to keep complexity (and therefore also energy requirements) to a minimum, but not to such an extent as not to store enough valid information.

As the apocryphal quotation from Albert Einstein goes: everything should be as simple as it can be, but not simpler. As the original says, observation of the environment will set the limit beyond which we must not simplify: "… the supreme goal of all theory is to make the irreducible basic elements as simple and as few as possible without having to surrender the adequate *representation of a single datum of experience*." (Einstein 1934).

## Sleep, Death and Reproduction

Because of the inevitable build-up of useless syntax, every intelligent system has three possibilities: learn more, forget what's useless, or die.

Learning means storing (and being able to process) new, valid information that effectively reduces uncertainty and, although energy is needed to store this new information in memory, being able to acquire new energy from the environment. Forgetting means freeing up memory from the syntax that's no longer used, one that requires energy without producing anything in exchange.

Forgetting is normally done through periods of reduced activity in which the system works on its internal reorganisation, processing less external information.

This is true for every complex system *a la* Herbert Simon, even non-intelligent ones. Some computer operating systems need to be periodically rebooted because they cannot manage the level of superfluous complexity they produce while operating. In order to postpone the need for rebooting, computers "reorganise themselves" during periods of scarce activity (as is the case with Microsoft Windows Disk Defragmenter).

Organisms with a central nervous system need sleep: while the role sleep plays isn't fully understood, it's widely accepted that one of its functions is to prune many of the synapses created while the organism is awake (Li et al. 2017).

Unfortunately, sleep, like the defragmenter, can however only put off the inevitable: death. In time, a computer will accumulate more and more useless syntax. The same goes for a car, a human being, or a civilisation. It's the natural order of things, and can be well understood in the information-energy framework. As some energy opportunities are exploited, they no longer arise again: for a lion it may be prey that no longer follows a certain trail to a water hole, for a company it may be that the market is no longer willing to pay a price that is higher than the cost of production. A lion lying in wait for prey that doesn't show, or a company manufacturing obsolete products, and storing the information necessary to do so, continues to require the same amount of energy as before, but with no energy return.

The history of management is full of stories of companies that, at the very peak of their success, collapsed disastrously because they couldn't shake-up the company and introduce new production processes. Kodak is a glaring example of this. Rebecca Henderson, one of the first scholars who studied corporate organisation to realise how important information and in-house organisation is for companies in order to evolve (Henderson and Clark 1990), said the following after Kodak went bankrupt: "Kodak is an example of a firm that was very much aware that they had to adapt, and spent a lot of money trying to do so, but ultimately failed." (Gustin 2012).

Interestingly, the company's senior vice president, had proposed the most sensibly solution: "Directing its skills in complex organic chemistry and high-speed coating toward other products involving complex materials" (Shih 2016). This would have meant not sacrificing the huge amount of *technical* know-how the company had accumulated over the years, but did require significant changes to the in-house organisation, which in the end proved impossible.

The last option was death. We see death as a tragedy, because most of the information stored by the organism seems lost to us. But this cannot be completely true: if information was irreversibly lost when an organism died, it would have been difficult for life to evolve. So there had to be a way to pass useful information on to future generations. The same is true for companies. The legacy of General Electric, which today is fighting to survive, in terms of management innovation, will remain in the "DNA" of many other companies for years to come. Xerox's research into the usability of computers paved the way for the success of companies like Apple and Microsoft.

Works of pure research stand out too: Claude Shannon was working for AT&T when he published his article on information theory in the company journal. Benoit Mandelbrot was working for IBM when he developed fractal mathematics, and today "Big Internet" business groups like Google publish fundamental studies on applied mathematics. Kodak left its legacy to human knowledge in the form of hundreds of scientific articles.

The same is true for individuals. There's a legend that tells of the religious conversion of John von Neumann before he died, refuted however by the scientist's brother (Hargittai 2008). Unlikely as it may seem, it is a plausible idea: how could the most intelligent mind of the time accept that all his knowledge would simply disappear the day he lost his battle with cancer? But the truth is that von Neumann's knowledge has not been lost, thanks to the thousands of people who have soaked up at least part of his intellectual legacy. Thousands of articles in IT, economics, physics and mathematics, have references to works of von Neumann.

Language and writing are *Homo sapiens'* solution to lost of information which would happen when people die.

Unicellular organisms learnt a similar trick billions of years ago. As mentioned above, DNA passes on information not only on how to construct the cell itself, a process perfected over hundreds of millions of years, but also on how to react in certain situations: something that can be learnt during a lifetime and passed on to descendants through epigenetic inheritance –"phenotypic variations that do not stem from variations in DNA base sequences are transmitted to subsequent generations of cells or organisms." (Jablonka and Raz 2009).

## The Limits of Unicellular Organisms

The DNA is rightfully seen as a milestone in the evolution of life. Thanks to it, unicellular organisms found a way to store information, and filter it through natural selection. More successful DNA strands were more likely to reproduce, less successful ones more likely to become extinct.

But cells soon reached the limits of this mechanism. In fact, the amount of information an isolated cell can store is limited by the fact that it cannot grow out of all proportion: in cells, nutritive material is only distributed by diffusion and the walls are made of protein chains that cannot extend over more than a certain surface area. For this and other reasons, the biggest unicellular organisms grow to just a few centimetres and these are very much a minority (Marshall et al. 2012).

A limit in size also limits information. To increase information capability, life stopped storing information in single cells – by now saturated – and started using networks. Although it occurred in different ways, first with amino acid networks and then with protein networks, life changed gear and introduced a new level of complexity. This didn't take long: traces of bacteria 3.5 billion years old show that there were communication's mechanisms even then (Decho et al. 2009), (Lyon 2015), (Zhang et al. 2012).

In recent decades the mechanism of bacteria communication, called *quorum sensing*, has revolutionised our idea of unicellular organisms: more than 99% of bacteria live in communities, so-called *biofilms*, where the organisms form "a complex web of symbiotic interactions" (Li and Tian 2012). Using quorum sensing, some bacteria (such as cholera) sense not only how many conspecifics are within signal range, but often also which and how many other species of bacteria they are sharing the environment with (Ryan, website), (Ng and Bassler 2009).

Many types of bacteria use quorum sensing mechanisms to do more than simple census taking. *Myxobacteria Xanthus* and *Amoeba Proteus* are examples of unicellular organisms that form colonies very similar to multicellular organisms: "In a process likened to 'the great animal herd migrations' up to a million cells move toward aggregation sites where fruiting bodies form. Of the initial cell population, only 10–20% will transform into long-lasting, stress-resistant myxospores and survive to reproduce another day. A staggering 65–90% of the initial population collectively suicide, by rupturing their cell envelope (autolysis). … The cell fragments are assumed to provide carbon and energy for development. Another 10% of the population transform into special cells that remain on the periphery … these cells could be a kind of sentry, to prevent pillaging of the sacrificial feast and predation of the dormant myxospores" (Lyon 2007).

## From Interaction to Cognitive Processes

We humans usually associate cognitive processes with just one lifeform: the one we call *sapiens* (Lyon 2015). Yet, as we've seen, biological networks like DNA have the ability to learn, and evolution favours the emergence of cognitive processes (Watson 2014).

To understand how simple mechanisms of interaction can produce cognitive abilities, researchers have studied various loose communities found in nature –those in which every element is linked to the others through just an exchange of simple signals. Similar interactions between elements regulate typically bacteria communities, flocks of birds and schools of fish (Deneubourg and Goss 1989).

In these cases the community learns through allelomimesis, or imitation, with the help of a mechanism called autocatalysis that prevents the indiscriminate dissemination of the behaviour of just one individual. "The probability of an individual adopting a particular behaviour is a function of the number of individuals already exhibiting that behaviour." (Goss and Deneubourg 1988).

For example, in a flock of birds on the ground, one bird takes flight because it notices a suspicious movement. The birds nearby lower their alarm threshold, and follow the first that took flight if they see even a slightly suspicious movement. At this point the whole flock takes flight, regardless of whether or not there is actually a predator. One bird acts as a trigger for a possible group behaviour through allelomimesis, autocatalysis induces the whole flock taking flight in case of danger.

Autocatalysis and allelomimesis allow the emergence of Hebbian learning processes, although rudimentary ones: "The strengthening of frequently used trails is also reminiscent of Hebbian reinforcement of active neuronal pathways through long term potentiation" (Couzin 2009).

They can often be observed in the behaviour of humans too. How does a peacefully protesting crowd turn into a crowd of looters? One protester becomes violent, and another might copy the behaviour (allelomimesis). But generally, all protesters have their own threshold for doing what others are doing. Each protester will become violent themselves only if a certain number of protesters have already started looting. Each new looter will increase then the probability of the crowd looting (autocatalysis) (Buchanan 2007).

Something similar happens when *fake news* goes viral on social networks: the logical-analytical abilities of the average user aren't comparable to those of experts. But the latter, who base their knowledge on experiments and cross-validation, typically don't dedicate much time to updating their *social* profiles.

Unfortunately for the quality of the information on so-called social networks, while scientific thought has only been around for 500 years, allelomimetic and autocatalytic behaviour has been part of how instincts have evolved for hundreds of millions – or even billions – of years. Therefore the comments of thousands of people who consider vaccines to be harmful, no matter how little they know about the subject, have more power to convince others than hundreds of scientific articles written on the same subject.

Facebook users act more like the lemmings of the mass suicide myth than a flock of herons[1]: when someone starts sounding off about vaccines, other users find it all but impossible to avoid following suit, creating a sort of domino effect.

Autocatalysis and allelomimesis of course weren't introduced by nature to favour the spread of *fake news* on Facebook. They were meant to guide the evolution of life towards forms of aggregation that were more, and not less, intelligent: it's thanks to these mechanisms for example that ants find the best and shortest paths to where they're going (Goss and Deneubourg 1988), a capability of insect colonies we will describe later in this chapter.

## Nervous System or the Forgotten Transition

As mentioned at the beginning of this chapter, the genetic code and the nervous system are two different memories – the first is long term, the second short term. The genetic code of living organisms lets them pass information on from one generation to the next, and do so for millions of years, and has changed the way we think about the evolution of life. The brain, on the contrary, loses the information it

---

[1] It's been proven that the only way to observe lemmings' mass suicide is through a Disney documentary: "The stampede of lemmings in Walt Disney's *White Wilderness* is an obvious fake" (Chitty 1996).

contains with the death of the organism. Probably for this reason, scholars often considered the role played by the brain as secondary.

In a fundamental work like *The Major Transitions of Evolution*, which focuses on the evolution of information processing by living organisms, John Maynard Smith and Eörs Szathmáry failed to consider the emergence of the nervous system as one of the fundamental transitions (Smith and Szathmary 1995).

In (Calcott and Sterelny 2011), Szathmáry himself acknowledges this slip: "the origin of the nervous system is a forgotten transition." He also adds that they would have included the nervous system in the second edition of the book.

From the point of view of processing information, and therefore that of this book, the emergence of the nervous system, up to the formation of the brain, is a fundamental stage in the evolution of life, and it's just as important as the emergence of DNA.

Jablonka and Lamb (2006), quoted by Szathmáry in the article above, write: "There are interesting similarities ... between the outcomes of the emergence of the nervous system and the transition to DNA and translation, in both of which the interpretation of information involves decoding processes."

From the first insects to the Anthropocene, the nervous system has had an extraordinary impact on the planet, transforming the interaction between life and the environment, as well as life itself.

In her autobiography, neurobiologist Rita Levi-Montalcini (1987) praises the imperfection of our species in relation to the environment. The fact that we are ill-equipped has forced us to continuously come up with new solutions – mostly produced by our brains.

The essential role of brains, with their ability to analyse data and react as a consequence, is "to serve as a buffer against environmental variation" (Allman 2000). The more sudden and unexpected these environmental variations are, the more organisms have to be able to adapt to them. And for this task, what could be better than short-term memory, able to see the patterns in everything around us?

## *Origin of Neurons*

As is often the case, neurons probably did not appear to do what they do now – create information-processing networks – but rather to find food: "Initially sensing environmental cues (such as the amino acid glutamate indicating prey) the partaking receptors and ion channels may have started to receive internal information (such as the transmitter glutamate) from within the newly evolving synapse" (Achim and Arendt 2014).

The role of glutamate has been studied in organisms that were probably the first to develop a nervous system, ctenophores, marine invertebrates also known as comb jellies, whose ancestors probably appeared towards the end of the Neoproterozoic Era around 550 million years ago. A ctenophore's nervous system coordinates its cilia; hundreds of tiny tentacles the animal uses to move around and search for food.

Unlike more complex nervous systems, the ctenophore's one mainly uses L-glutamate as its neurotransmitter (Moroz et al. 2014), which leads us to consider the hypothesis of the origin of neuron transmission (Achim and Arendt 2014).

Of a ctenophore's many properties, the most interesting could in fact be that it appeared before sponges. In other words, the first complex organisms to form might have adopted immediately a rudimentary nervous system, unlike sponges (Shen et al. 2017).

This has caused quite a stir. Firstly because it's not universally accepted as having been proven (Marlow and Arendt 2014), and secondly because it goes against the idea – officially denied but intuitively felt – that evolution is always a process of complexification.

The paleo-geneticists will have to make up their minds about the first point, but the second appears to be completely unjustified. Evolution means making the process of storing useful information more efficient, also reducing the amount of energy required to store it (using lighter balls as in the example at the beginning of this chapter).

Going from ctenophores, more complex and active, to sponges, simpler and with modest energy requirements, is just as much a winning strategy as making the nervous system more powerful and finding new solutions to the resulting energy crisis.

Having said this, how could we fail to agree with Rita Levi-Montalcini? Our conscious organisms, yearning for knowledge, with memories of past life – all made possible by our brain – seems a more interesting condition than the static existence of the humble sponge. We should consider ourselves lucky that evolution took a more complex path that led to *Homo sapiens*.

## First Brains and Shallow Neural Networks

We humans have got to know our brains so well, now we know that all we know is all but nothing. Nonetheless, by examining simpler natural neural networks and artificial ones, it's possible to come up with some hypotheses as to how information processing capability evolved in time.

The only brain that's been mapped until now is that of the hermaphrodite and the male *C. elegans*. The connectome of the hermaphrodite's brain (White et al. 1986), with its 302 neurons, was compiled in 1986 and has been studied extensively, as can be seen in (Watts and Strogatz 1998). The connectome of the male, with 383 neurons, was compiled, in 2012 (Jarrell et al. 2012).

The difference between the nervous systems in the two genera is due to the fact that the male *C. elegans*, which cannot inseminate itself, has a further purpose in life in addition to energy provision: to mate with the hermaphrodite.

As every adolescent knows, the cognitive processes behind mating mechanisms are challenging. Considering that the *C. elegans* mating ritual isn't as simple as the animal's position on the evolutionary tree might lead us to believe (Jarrell et al. 2012), it's almost a miracle that the male *C. elegans* manages to do its job with just a sprinkling of neurons.

Basically, *C. elegans* has developed a shallow neural network,[2] in which the sensory layer is a Hopfield network able to recognise some categories, called *Gestalt* by Hopfield (1982), of the surrounding environment – for example if there's a hermaphrodite in the vicinity, and what position it's in. After this initial classification, the information is sent to a second layer (interneurons) and then to a third, last layer, the motor neurons, which pass the stimulus to the relevant muscles (Varshney et al. 2011).

In the same way as in so-called Recurrent Artificial Neural Networks, some neurons have a processing memory and re-analyse their own output.

Considering that nematodes (the *phylum Elegans* belongs to) appeared soon after ctenophores (Poinar 2011), it didn't take long for evolution to go from a distributed neuron network that could only move cilia, to one organised in a way that could process information and produce relatively complex behaviour.

By forcing the analogies with artificial networks a little, we've gone from the Hopfield Networks of 1982, to Recurrent Neural Networks, developed during the 1990s, in a few million years.

## Societies and Natural Selection

If natural selection were the only mechanism of evolution, one might think that communication between complex organisms would not have gone beyond reproduction. Why collaborate with other individuals? This would mean increasing the possibilities of success of the *other*, with *their* genetic code, which isn't exactly advantageous to *our own* genetic code. Or even better, why allow other individuals' genetic code to propagate at all, and invent instead a way to clone itself?

The idea of natural selection influenced more than biologists. As seen in Darwin's *The Descent of Man*, welfare and redistribution of wealth were thought to play against the fitness of the species: the weak must perish to leave room for the strong. Scientists didn't even think collaboration was possible among animals. Therefore, they would never say that a collaborative society could be more successful than a non-collaborative one: collaboration was "human misbehaviour".

Ronald Fisher (1930), one of the forefathers of statistical experimentation, in *The Genetical Theory of Natural Selection* considers an original hypothesis on how human civilisations first formed. At a certain point in history, some primitive populations entered a "barbarian" phase. Barbarian societies were organised to evolve through selection: they were divided into social classes, with those best suited to survival at the top, and the strongest elements were selected through family feuds. The strongest were rewarded with reproductive prizes (e.g. polygamy), and finally, they adopted a lineage cult, to favour the male offspring of the strongest.

---

[2] Networks with just a few layers are called "Shallow Neural Networks", the opposite of "Deep Neural Networks" – the current superstars, with dozens of layers.

These barbarian mechanisms, according to Fisher, resulted in the "social promotion of fertility into the superior social strata". Human societies, and only human ones, started to favour the process of selection: "The combination of conditions which allows of the utilization [sic] of differential fertility for the acceleration of evolutionary changes, either progressive or destructive, seems to be peculiar to man." (Fisher 1930)

As had been the case for life according to Dawkins (2013), suddenly a structure appeared that could evolve thanks to selection – but a mechanism that explains the emergence of this structure isn't introduced (see the analysis of Johnson and Sheung Kwan Lam 2010). Dawkins explains the origin of life with the sudden appearance of the Replicator, which can eventually evolve through natural selection. Fisher is no different: suddenly, the society appears, so that can evolve through natural selection. Neither Dawkins nor Fisher suggest mechanisms that explain why and how these structures emerge.

Never mind that, if societies appear to let natural selection do its job, it becomes difficult to explain the emergence of fully collaborative societies (bees and ants), where no internal selection takes place.

The idea of emerging properties, "the whole is more than the sum of the parts", was still a long way off. It seemed necessary to imagine a mechanism that made it possible to apply natural selection to the evolution of communities of organisms.

The mathematics of genetic variation within populations under the effects of natural selection was eventually perfected by John B. S. Haldane (1932) and finally William Hamilton (1964).

In his *inclusive fitness theory*, later called *kin selection* by John Maynard Smith, Hamilton explains that when we behave altruistically with a relative with whom we have a certain amount of genes in common, we do so if the person we are being altruistic with shares some of the same genes.

As an example, if your sister steals €100 from you for an investment that makes €250, you've both made a profit. 100% of your genes have lost €100, but 50% of your genes –the one you have in common with your sister – have earned €250. That means on average your genes have made a €25 profit. The fact that the champagne eventually purchased with the €250 gain doesn't whet *your* taste buds, is but a minor detail: the whetted taste buds have 50% of your genes, so you reap the benefits in any case.

In *The Selfish Gene*, Dawkins mentioned it would be worth studying nine-banded armadillos, the female of which gives birth to a litter of identical quadruplets: "some strong altruism is definitely to be expected, and it would be well worth somebody's while going out to South America to have a look" (Dawkins 2013).

As can be expected, Dawkins' guess was disproven: despite the fact that armadillos can tell the difference between their twins and non-twins, they apparently behave in a friendly way towards all, regardless of shared genes (Loughry et al. 1998).[3]

---

[3] If you wonder why the quadruplets, it appears that the particular shape of the uterus causes poly-embryony. It's nothing to do with kin-selection (Loughry et al. 1998).

What Blanche says at the end of Tennessee Williams' *A Streetcar Named Desire* goes for armadillos too: "I have always depended on the kindness of strangers".

In fact, "Hamilton's rule almost never holds" (Nowak et al. 2012).[4] The most extreme form of collaboration, eusociality, would appear to be exactly what's needed to refute Hamilton's rule. In eusocial societies "adult members are divided into reproductive and (partially) non-reproductive castes and the latter care for the young" (Nowak et al. 2012). Eusociality is more common in societies in which siblings have 50% of their genes in common with each other than in those with 75%, as is the case with some bees and ants.[5]

Finally, the behaviour of individuals appears to be collaborative in many societies, regardless of how many genes the individuals have in common with each other. Furthermore, no animal societies – as Fisher mentioned – have established an internal selection system.

We must conclude that, from the point of view of evolution, organisms that collaborate for the common good have a greater chance of surviving than those that don't. This is just as true for humans as it is for other animals. Actually, more so for humans.

Internal selection, as Ludwig von Mises mentioned shortly after the fall of fascist regimes in Italy and Germany, seems to be there to favour tyrants:

> As every supporter of economic planning aims at the execution of his own plan only, so every advocate of eugenic planning aims at the execution of his own plan and wants himself to act as the breeder of human stock (Mises 1951).

## Insects and Intelligent Societies

The first organised societies of complex organisms emerged 300 million years ago in insect communities, in the form of eusociality in fact.

Without the help of parental selection it would appear to be difficult to explain why certain insects chose to help each other, sacrificing their own fertility: if they had an "altruism gene", they would favour the survival of others more than their own, so natural selection could not propagate the "altruistic gene". By definition, genes that don't follow Hamilton's rule, like the altruism gene, should become extinct.

---

[4] We could examine Hamilton's rule considering how much DNA (rather than how many genes) we have in common with other living beings. All humans have practically the same DNA in common, so altruism should be mandatory, total and undiscerning for all. We should expect extremely altruistic behaviour also from chimpanzees and towards them too, as we have 98.4% of our DNA in common with them. Considered in these terms, Hamilton's rule doesn't work, not because we're too altruistic, but because we aren't altruistic enough!

[5] In haploid-diploid sex determination, males hatch from non-fertilised eggs and therefore always inseminate the queen with the same genetic code: in this way, sisters on average have 75% of their genes in common with each other because the father's genetic code is always the same.

But ".. eusociality is not a marginal phenomenon in the living world. The bio-mass of ants alone composes more than half that of all insects and exceeds that of all terrestrial non-human vertebrates combined. Humans, which can be loosely characterized as eusocial, are dominant among the land vertebrates" (Nowak et al. 2012).

Natural selection can prevent unsuitable life forms from developing, and favour the success of a mutant organism. But it cannot explain either the emergence of life, or the change from unicellular organisms to complex organisms, nor the organisa-tion of the latter into societies.

The fact that natural selection can filter behaviour that's unsuitable for survival doesn't imply that every new behaviour emerges only thanks to selection. If A implies B, B doesn't necessarily imply A: behaviour that's better suited to survival is favoured by selection, but this doesn't mean that if a new behaviour emerges the cause is necessarily natural selection.

The essence of an intelligent system is to continually organise itself in a better way to absorb more energy, storing more and more useful information. The concept of *evolvability* is more important than that of *fitness*: the first is a long-term invest-ment, which goes beyond the current generation and genes, the second only lets behaviour emerge that increases the probability of survival of the *individual, now*.

An intelligent system that doesn't implement a long-term survival strategy is not worthy of the name. Improving cognitive capacities, or creating new (perhaps exter-nal) ones would appear to be an excellent strategy. This is valid for all intelligence systems, from the biological cell to the nation-state.

The creation of networks to increase the amount of available information and the capacity to process said information is a common phenomenon. For data analysts, and high energy physicists, the use of personal computers in the 1990s was severely limited by one factor: the capacity of local hard disks. At CERN, over a period of 5 years, I personally witnessed the change-over from workstations, kind of powerful PCs, with their huge external hard disks, to consumer PCs, which were cheaper, could process data stored in shared file systems in parallel (NFS and AFS, similar to virtual disks) and distributed databases. A path which culminated with the Web, giv-ing non-professional users access to a virtually infinite amount of information, incomparable with what's on the hard disk of their PC.

This process has intensified in time: today, a typical PC (smartphone) can store just a few tens of gigabytes on the hard disk (not much more than what I had on my workstation in 1998), but processes – in the case of adolescents – hundreds of giga-bytes every week[6] (approximately the amount of data I processed during my PhD in the 1990s).

Today, having the same storage system shared by millions of people, gives the *"Homo smartphonicus"* access to an abundance of information, something that would have been unimaginable until recently, with only-local storage.

---

[6] When kids go back to school in Italy there's a 30% drop in nationwide 4G data traffic (personal communication, Wind Italia).

Something similar happened with eusocial insects, the first animals to develop societies.

Insects were the first to reach an information-processing limit, for the same reason as biological cells a few billion years earlier. Although giant insects roamed the earth during the Permo-Carboniferous period,[7] 300 million years ago, is not convenient for insects to grow into giants. Quite the contrary: they've implemented various miniaturisation strategies as they evolved. Their compact dimensions have been essential for "microhabitat colonization, [and the] acquisition of a parasitic mode of life, or reduced developmental time as a result of a rapidly changing environment" (Niven and Farris 2012).

To sustain and perfect miniaturisation, insects have developed a passive respiratory system, consisting of spiracles and tracheae that let the cells exchange carbon dioxide for oxygen. As a result, the nervous system developed in a decentralised way,[8] providing a greater surface area to oxygenate in terms of total volume[9] (Niven and Farris 2012).

This miniaturisation strategy comes at a cost then: a cognitive limit of the individual, due to the concentration of oxygen. Processing information no longer as individuals, but as a network, increased then the information processing capacity of the species, without renouncing to miniaturisation.

Ants, for instance, communicate with dozens of different pheromones (Jackson and Ratnieks 2006): each ant leaves a track as it passes that acts as a signal for the next ant that passes that way, and each message is subsequently reinforced, weakened or changed by the other sisters in the colony.

Using this relatively simple mechanism, ants, which are blind, successfully solve "the travelling salesman problem": Given a list of cities and the distances between each pair of cities, what is the shortest possible route that visits each city and returns to the origin city? (Yogeesha and Pujeri 2014).

It sounds like a miracle, but ants really can find the fastest route between a certain number of points. As the problem is one of the toughest in mathematics, for which no one (including ants) have yet found a completely satisfactory solution (in other words valid for any number of cities), it is a noteworthy result.[10]

---

[7] There are fossils of dragonflies with a wingspan of 70 cm and millipedes two metres long. In addition, insects exposed to an artificial atmosphere with a high oxygen content develop (through the activation of some genes) hypertrophic features (Zhao et al. 2010). The fact that "gigantism" in insects only occurs in hyperoxygenated atmospheres is worthy of note: it's not accidental. Organisms able to activate genes that develop a phenotype that's more suitable to survival are selected: selection acts as a filter.

[8] For centuries we have known that an insect can survive for weeks after the part of the nervous system in the head has been removed (Gregory 1763)

[9] Two "spherical" brains have a surface area 25% larger than one brain of the same volume

[10] The algorithm used by ants is… as simple as it can be, but not simpler. At first, as they're blind, they move randomly. In a certain period of time the ants that chose the fastest path with have travelled back and forth several times, reinforcing the pheromone trails. After a while, the most distinct trail will be the fastest (Dorigo and Gambardella 1997).

As was the case when unicellular organisms organised themselves into complex organisms, the emergence of sociality in insects as a mechanism used to store and process information has not certainly been the dominant paradigm in science, but it is present.

First of all, the idea that information processing capabilities can spontaneously emerge is deeply-rooted in information technology and neuroscience:

> The bridge between simple circuits and the complex computational properties of higher nervous systems may be the *spontaneous emergence* of new computational capabilities from the collective behaviour of large numbers of simple processing elements (Hopfield 1982).

In addition to that, for other geneticists, "gene-based theories have provided plausible, albeit incomplete, explanations [for the origins of collaborative societies] ... In many social groups, socially learned cooperative behaviours increase the productivity of the individuals in the group relative to that of animals that are not group members, yet the evolutionary effects of such behaviourally transmitted information have rarely been explored." (Jablonka 2006).

There is a long way to go from ants colony solving the salesman problem to *Homo sapiens* inventing GPS satellite system and creating car navigators. But the road is marked out.

## The Social Body

To many biologists, including Bert Hölldobler and Edward Wilson (2009), the authors of *The Superorganism: The Beauty, Elegance, and Strangeness of Insect Societies*, the problem lies in understanding when and how the "eusociality gene" became established. The reasoning behind this is that, if a complex organism is altruist, this is encoded in the information code of each individual.

But we are dealing with complex systems: a property of the system doesn't derive from a property possessed by each individual – this is in fact Simon's definition of a complex system. Cooperation can be one of the many effects of communication: first comes the ability to communicate, then the instinct to cooperate for a common scope.

If we consider complex organisms, including human beings, this hypothesis doesn't appear to be too far-fetched. For human beings, cooperation without prior communication is something of a rarity. Many anthropologists who have lived with hunter-gatherer populations, say that learning the language was essential not only to their studies, but for their safety too. When Daniel Everett was with the Piraha he acknowledged that on one occasion he only managed to avoid being killed by a drunk member of the tribe because he was able to speak the man's language (Everett 2009).

Even Frans de Waal, a primatologist famous for his idea that culture and morals are not exclusively human traits, acknowledges the fundamental role played by language in the formation of an ethical society:

> Members of some species may reach tacit consensus about what kind of behaviour to tolerate or inhibit in their midst, but without language the principles behind such decisions cannot be conceptualized, let alone debated. To communicate intentions and feelings is one thing; to clarify what is right, and why, and what is wrong, and why, is quite something else (Waal 1996).

Surely chimpanzees communicate, and do form collaborative societies. But nothing compares to humans, and in fact there's an incredibly high level of collaboration between humans "What is most amazing is that our species is able to survive in cities at all, and how relatively rare violence is." (Waal 1996). No other primate could live in cities as crowded as ours without starting a civil war.

For Robin Dunbar, the anthropologist famous for proposing that group size was correlated with brain size among species of social primates, language lays the foundations of society. What's more, according to Dunbar, language, in the form of gossip, emerged not as a form of communication, but as the evolution of *grooming*: "gossiping ... is the core of human social relationships, indeed of society itself. Without gossip, there would be no society." (Dunbar 2004).

As mentioned in (Reader 1998), language isn't much use when hunting. There's really no need to read the adventures of anthropologists gone native, from Colin Turnbull in *The Forest People* (Turnbull 1961), to Daniel Everett, who tell of their inability to move through the forest in perfect silence on a hunting expedition. As every hunter knowns (Forense 1862), language is not only superfluous, it's to be avoided at all costs: vocal skills are used just for "faking natural sounds in order to lure their prey within range of their weapons" (Botha 2009).

But language does have something to say when it comes to hunting. Language in fact is fundamental if hunters are to peacefully share out game: it therefore emerges more as a means of negotiation (Reader 1998), rather than just gossip, or a way to pass on information.

This goes for every means of communication. While communication, to be efficient, has to pass on "useful" information, this does not necessarily have to be the case. The exabytes of cute kittens – hardly informative – sent over Facebook might have a role: communication is also used to create the system's identity. The role of language and communication in general is also to make individuals feel – and therefore become – part of the system as a whole.

Language permits the constitution of the social *body*, a structure where the single elements go beyond collaboration. In Jean Jacques Rousseau's words means "the *total alienation* of each associate, together with all his rights, to the whole community" (Rousseau 1762). We call it *body* for a reason: the citizens become the cells of an organism, they stop existing as a single element. When we raise an arm to defend our head from a dangerous object, there's no altruism in the arm's action: the cells of the arm wouldn't survive outside the organism anyway: the *self* of the cells in the human body is a meta-self, transcendent, not immanent.

For Rousseau, in human societies, like in insect societies, the individual ceases to exist, and is alienated mainly thanks to shrewd communication – the conviction that every person has a well-defined role to play in society, and must play that role and no other.

Language allows the emergence of collaborative societies, but not necessarily fair societies –exactly like ant colonies. Citizens under the social contract (which is an emergent property in Rousseau writing, not an actual piece of paper) are safer. But certainly not more free or smarter than they were.

# Chapter 4
# The Human Social Brains

We saw how the emergence of biological neural networks, from the simple neurons of ctenophores to *C. elegans'* neural network, appears to be a natural evolutionary process.

In this chapter, however, we'll consider one brain in particular – our own. What drove the development of our hypertrophic neural network?

According to some scientists, there is no answer to this question. Yuval Harari (2014) writes: "What then drove forward the evolution of the massive human brain during those 2 million years? Frankly, we don't know."

A good approach would be to consider the development of our brain under the same light we would consider the development of other brains. In the framework of intelligent systems, it comes natural to consider the evolution of the communication skills of mammals in general, primates, and *H. sapiens* in particular, as a transfer of the cognitive abilities of the species from individuals to the network, just as it happened for unicellular organisms, which are networks of proteins, and complex organisms, which are networks of cells.

The leitmotif of this book is that there wasn't just one single cognitive revolution – typically identified with human language – but rather a continuous increase in the capability of intelligent systems to process and store information. And the revolutions, if we really want to identify any, are the emergence of new forms of life based on the integration of old form of life.

Revolution, therefore, can be seen as starting with the introduction of culture, or information stored by the society and passed on from one generation to the next. And culture, in this sense, was not invented by *H. sapiens* 50,000 years ago. Not even by some of our ancestors 14 million earlier ("great-ape cultures exist, and may have done so for at least 14 million years," van Schaik et al. 2003). The need for a cultural system is tied to the inevitability of death and the need to reproduce.

As mentioned in Chap. 3, death and reproduction are necessary processes that clean superfluous syntax. But a system that was unable to pass on the useful information acquired from one mortal individual to other surviving individuals wouldn't be a great evolutionary success. Intelligent systems have to adopt a mechanism to

© The Author(s) 2020
M. Alemi, *The Amazing Journey of Reason*, SpringerBriefs in Computer
Science, https://doi.org/10.1007/978-3-030-25962-4_4

pass on useful information to other systems before they die: this is the only way these systems can evolve into ones that are more and more intelligent.

It's true that today we know organisms can pass on information to their descendants by activating and deactivating genes, but as mentioned above, DNA is a long-term memory, difficult to modify. What's more, it can only be used to pass on information to one's descendants. It is a one-to-very-few communication system.

Culture on the other hand is the capability to store information not in an individual, but in a network. And the greater the communications' capability of the elements in a network, the more information the network can store and process.

So, intelligent systems like uni- or multicellular organisms have two mechanisms for recording new survival strategies over multiple generations: as an individual by adapting the phenotype[1] through genetic modification ("… novel phenotypes arise as a result of environmental induction". Jones 2012), and passing this information to their offspring; as a network by modifying its structure, for instance through Hebbian learning.

In the case of complex organisms, we know for a fact that some mammals adopted social behaviour –an exchange of information between organisms– as far back as the Palaeocene, immediately after the Cretaceous period, the age identified with the extinction of non-avian dinosaurs 65 million years ago. The remains of *Pucadelphys andinus*, a small Palaeocene marsupial mammal, suggest they were polygynous animals with a highly evolved social life living in a pack with the males probably competing with each other (Ladevèze et al. 2011).

We don't know when these behavioural patterns emerged. But essentially, along with social behaviour, at that time mammals had already developed a brain that, when compared to the brain of a dinosaur, is like comparing a Ferrari Testarossa to a Fiat 500.

Although we studied the emergence of the nervous system, we didn't consider how difficult it is, for an organism as a whole, to maintain a brain even as simple as that of Palaeocene mammals, and make it work well. Just as the sophistication of a Ferrari depends on more than just a powerful engine, a mammal's brain also needs just as sophisticated an organism to go with it.

First and foremost, a powerful brain tends to also become an information processing centre, and therefore a *single point of failure*.[2] All it takes to put it out of action is a relatively light shock, so the brain needs a strong and costly skull to protect it. What's more, as is the case with every powerful motor, it requires a notable amount of energy (in the case of *Homo sapiens* and some dolphins (Martin 1996) it weighs just 2% of the body weight but requires 20% of the energy absorbed by the organism).

In addition, energy is required continuously, because the brain, unlike muscles, cannot store nutritive substances. If the brain is left without nutrients and oxygen

---

[1] Phenotype: "The set of observable characteristics of an individual resulting from the interaction of its genotype with the environment." (Oxford English Dictionary)

[2] A part of a system that, if it fails, will stop the entire system from working (https://en.wikipedia.org/wiki/Single_point_of_failure, verified 5 May 2019)

for just a few minutes, it suffers irremediable damage, exactly like modern data centres would be damaged if they were suddenly deprived of electrical power.

Also like data centres, because of the amount of energy consumed, our brain heats up – something that poses problems in the design of the skull and circulatory system. The very definition of "mammal" includes the presence of a cerebral neo-cortex and a particularly well-developed cerebral cortex (Borrell and Calegari 2014). Both these areas can only function thanks to a sophisticated body temperature regulating mechanism, which birds, reptiles, fish –and of course dinosaurs– don't have.

In short, in terms of a cognitive revolution, the real revolutionary leaders were some obscure vertebrates who started investing in the cognitive organ some 200 million years ago: "What is beyond dispute is that the earliest mammals themselves did have significantly enlarged brains … brain size in Mesozoic [248 million to 65 million years ago] mammals lay within the lower part of the size range of the brains of living mammals. This represents an overall increase of some four or more times the volume of basal amniote brains, and presumably involved the evolution of the neo-cortex, the complex, six-layered surface of the cerebral hemispheres that is one of *the most striking of all mammalian characters*" (Kemps 2005).

## Why a More Powerful Brain?

Understanding the origin of the mammals' brain is fundamental if we want to understand the origin of our own. According to some anthropologists, our brain has only really been useful in the last few thousand years: "For more than 2 million years, human neural networks kept growing and growing, but apart from some flint knives and pointed sticks, humans had precious little to show for it." (Harari 2014).

Looking at it from this point of view, the brain appears to be a mere exercise in sexual energy-wasting (Miller 1998), which by chance, and only in the case of *Homo sapiens*, around 50,000 years ago produced symbolic thought. Just like birds that grew wings for no obvious reason, and then by chance found they could fly.

But the brain is a costly investment. According to Wheeler and Aiello's (1995) *expensive-brain hypothesis*, many complex organisms, in order to afford an increasingly more powerful brain, have in fact reduced, as they have evolved, the mass of other energy-consuming tissues like the digestive system and muscles. *Homo sapiens*, with their thin body and short intestine, is an excellent example.

If the brain was so useless as Harari wrote, this investment would appear to make no sense. A useless, costly brain, that needs so much energy it forces an organism to reduce organs essential for its survival like muscles and the digestive system cannot have laughed in the face of natural selection for hundreds of millions of years.

If a mutation, whatever the cause, is prejudicial to *fitness*, it's hard to imagine why it would be adopted by the entire species.

To justify the introduction of the brain we can imagine a few possibilities.

First, it could be that the brain initially develops like a parasite, to the detriment of its host organism. The brain "decides" it will have the host organism mate only with other similarly hyper-cerebral conspecifics.

In this case, a useless brain – the one indicated by Harari – will come out on top anyway, in spite of the process of selection: any organism able to increase its cerebral capacity to the detriment of other organs will be taken down this path by its parasite-brain.

This may be the case, but perhaps the brain evolved for a more noble cause, like its cognitive abilities. The brain's problem-solving capabilities could represent a profitable return on the investment. In fact, for many years "it was assumed that brains evolved to deal with essentially ecological problem-solving tasks" (Dunbar 1998): it's cold, so instead of waiting for selection to favour hairy people the brain invents clothing.

Another evolutionary push, not necessarily an alternative to the one above, could be Dunbar's *social-brain hypothesis*. Dunbar shows that the evolutionary push towards a more powerful brain in primates derives not from the need for cognitive abilities, but from more sophisticated *social* skills. More and more organised *Homo* societies made this genus a success, despite the atrophy of the organs.

Dunbar's hypothesis could, and perhaps should, be extended to include mammals, the first organisms that made a serious investment in the brain: "human culture cannot be disassociated from social life, and therefore from humanity's *mammalian* and primates foundation" (Sussman 2017).

So, getting back to the reason why mammals in general, and primates and *Homo* in particular, have such a powerful brain, we have three possible hypotheses:

1. The brain of mammals, like an internal parasite, developed to the detriment of other organs, even though this wasn't really necessary.
2. Brains evolved to become more and more powerful to improve *problem-solving* capabilities.
3. Same as above, but in this case to create more complex societies.

Obviously, none of the three hypotheses can be discarded completely, but the social component, the third, appears to have carried more weight than the other two. Isler and Van Schaik (2008) analysis on mammals and modern dinosaurs – birds – shows that the brain, when unable to favour collaboration, is more of a liability than an asset.

In practice, the body needs to reduce its energy requirements when the whole organism cannot exploit the advantages that derive from the ability to collaborate with others of its species. This means for instance reducing the digestive system in mammals or the pectoral muscles in birds.

The "maximum rate of population increase[3] is negatively correlated with brain size only in precocials [offspring born relatively mature and mobile, e.g. ducks] and

---

[3] The *maximum intrinsic rate of increase* is the per capita birth rate minus the per capita death rate for a population (Cole 1954).

semi-precocials, but not in altricials or semi-altricials [offspring requiring total care, like eagles]" (Isler and Van Schaik 2008). In other words, birds with a high encephalization level but a scarce ability to pass information on from parent to offspring, reproduce *less* than those that, with a brain of the same power, create a parent-offspring bond.

Note that Isler and Van Schaik do not think their study contradicts the expensive-tissue hypothesis:

> The observed trade-off between the maximum rate of population increase and both absolute and relative brain size supports the notion that this trade-off is caused by an energetic constraint, especially since it disappears in lineages where the mother's energetic burden during reproduction is alleviated through helpers. Thus, our results fully support the expensive brain hypothesis, which predicts that relatively large brains can evolve only when either energy input increases. (Isler and Van Schaik 2009)

In practice, if energy is available, organisms, and rightly so, will develop their brain and pick the low-hanging fruit that derives from problem-solving skills. But "during mass extinctions large-brained taxa are especially vulnerable," and precocial animals, considering the same cerebral energy consumption, are more vulnerable than altricial animals:

> But what change of lifestyle would allow the evolution of larger brained lineages? Our results show that, as predicted by the expensive brain hypothesis, allomaternal energy inputs during offspring production are one critically important factor. (Isler and Van Schaik 2009)

In a similar analysis, this time focused on mammals and primates, Isler and Van Schaik (2012) proved that the brain of the genus *Homo*, without a significant capability for collaboration, would already have been unsustainable already 1.8 million years ago, when the successful *Homo erectus* started colonising the whole planet.

In short, Fisher's idea – that societies formed so natural selection could improve the species – could not be further from the truth. Societies, made up of individuals with a communicative and collaborative brain, are an essential instrument in order to be able to afford even more powerful brains. Mammals' brain can afford to be so powerful precisely because it's so spectacularly collaborative.

Human brain in particular appears all the more extraordinary not for its calculation capabilities, but for its communication capabilities. Just as neurons probably first appeared as intelligent cells, able to report the presence of food, then showing their "true colours" when they learnt to communicate, forming neural networks, the brain in turn was born to be an information processing centre that then went on to become the key to success of some species – *Homo sapiens* first and foremost – when it learned to communicate.

This success is evident also in consideration of the biomass on the planet for different organisms, shown in Fig. 4.1

Even excluding *Homo sapiens* and the animals humans have domesticated, the biomass of mammals is three times that of surviving dinosaurs (the birds). Mammals

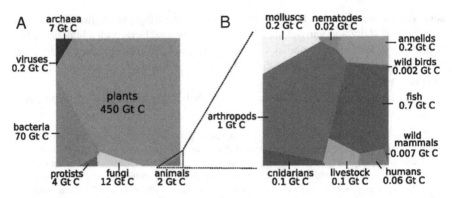

**Fig. 4.1** Graphical representation of the global biomass distribution by taxa (Bar-On et al. 2018)

today, with *Homo sapiens* leading the way in a *de facto* dictatorship, represent the vast majority of terrestrial vertebrates.[4]

The success of mammals is the success of an evolutionary strategy aimed at developing an organ that doesn't merely excel in processing information: it does a great job of exchanging information too, and therefore creating networks that can quickly store and process information necessary for survival.

## The Primates' Brain

Isler and Van Schaik (2008) analysis of birds appears to be convincing, and Isler and Van Schaik (2012) extends these conclusions to primates. The importance of their studies can hardly be overestimated. Until the last century, anthropology considered the social behaviour of primates to be less important than the creation and use of tools. The number of results for "primate social behaviour anthropology" and "primate tool use anthropology" on Google Scholar shows the complete disinterest of the post-war generation in the social relations of primates compared to the use of tools: 800 compared to 17,000 hits.[5]

But in time, around the year 2000, articles on social relations surpass those on the use of tools. Today, the figures are 90,000 for social relations compared to 50,000 for tools. Dunbar's *social brain hypothesis* was published in 1998.

Ape the tool-maker and ape the social animal are representative of the second and third hypothesis for the emergence of the brain – problem solver or social instrument.

It makes sense that paleoanthropologists gave initially more weigh to the tool-maker. The reason for making tools is evident: an increase in energy gain.

---

[4] There are less vertebrates than arthropods, such as insects, spiders and crustaceans, however.

[5] Analysis carries out with Google Scholar, 27 february 2019.

For both man and chimps this is certainly true. *Homo habilis* fed on marrow by cutting bones open with flint tools, and chimpanzees obtain high calorie foods such as honey, termites, marrow and brain from their prey by using specific tools to do the job.

But is not as simple as it looks. Simple energy gain cannot, for example, explain the inventiveness of the gorilla. Far from stupid, the gorilla doesn't (as far as we know) use any tools to increase the amount of food it has access to (for that matter it has plenty of leaves to feed on). A gorilla does, for example, create tools to help it move through water (Breuer et al. 2005). Millions of years of evolution to develop a brain that weighs half a kilo with its relative energy requirements that only comes up a stick to help you cross a stream isn't exactly an excellent investment.

But the brain did not give gorillas just that. The mammals' brain, in general, performs *two* tasks. The first is to find solutions to the problems found, a task often performed by one or a few individuals. The second is the cultural absorption of the information acquired, in other words storing information in the same society using communications mechanisms like imitation.

Gorilla didn't develop their brain to invent a stick to cross a stream, but to communicate: some populations communicate using rudimentary sign language (Kalan and Rainey 2009).

Although archaeology was established specifically to study the use of tools by primates (Haslam et al. 2009), it emphasises the importance of conformist transmission mechanisms – in practice learning through imitation or allelomimesis as mentioned in Chap. 2 (Luncz et al. 2015). In a sense, the invention of the tools by the brain is the finger pointing at the moon, but the moon is the society.

In conclusion, as is the case for mammals in general, the primates' brain is first and foremost a social brain, and has been for a long time. The ability to live in communities made it possible to store information not only as individuals, but as a society. Individuals die and lose information, societies don't, or at least not so fast.

Something similar happens with ants, although while ants have developed a sort of distributed brain and the information in the same dies with the swarm, mammals, and primates in particular, have created a network that can not only share information quickly within the same, but also keep it for millions of years.

Although on a different scale, this isn't introducing anything new compared to what was described for unicellular, multicellular organisms and insect superorganisms in the previous chapters.

## What Makes a Homo

Paraphrasing "The Big Lebowski" (Coen and Coen 2009), the question "What makes a *Homo*, Mr Lebowski?" has no easy answer. As in the sorites paradox (or paradox of the heap), in which we don't know when to start calling "a few grains of sand" a "heap of sand" but do recognise the two as being different, in palaeoanthropology it's hard to define exactly when the brain of the ape lost ground to that of *Homo*.[6]

But today it's easy to see the difference between the two brains: *Homo sapiens* brain is the one that has by far come up with the most complex tools *and* societies.

While mammals are the organisms that have invested the most in the brain in order to communicate, and therefore store and process information in a network of organisms, *Homo sapiens* is the mammal par excellence: the one that has taken its ability to communicate to new limits, and is still pushing those limits.

The truth is, a martian anthropologist would probably call ourselves *Gens communicans*, and not *Homo sapiens*. *Homo,* more than any other complex organism, continues to aggregate into complex structures thanks to new communication media, creating a global network of people – a single tribe, *gens*. We are an organism that started to create a new level of aggregation: that of the meta-organism, an organism made up of complex organisms.

So, instead of attempting to define the moment in which the genus *Homo* appeared, a problem that's as paradoxical as a heap, we can try to analyse the process that's made us what we are. In other words, a species with some apparently unique characteristics, such as highly symbolic thought and the use of extremely complex languages, with which we can send highly informative messages in an efficient way.

And as in the sorite paradox, there won't be the first *Homo*, but instead a continuous evolution from hardly communicative (for our own standards) apes to hyper-communicative *Homo*. We need therefore to analyse the emergence and evolution of language.

Above, we saw that a more and more demanding brain developed into a brain that was not only able to autonomously solve certain problems, but also communicate. This is because the ability to connect with other individuals means being able to store and process information as a network. And being part of a communication network means being pushed into communicating even better.

It seems improbable that the changes our body had to go through in order to develop the use of spoken language all happened by chance (Lieberman 2014). Not only a more powerful brain was necessary. The tongue, the soft palate and the glottis underwent major transformations which, together, made it possible for *Homo sapiens* to speak. As each transformation requires the activation of various genes, we cannot say there's one single "language gene".

Again, considering life as the evolution of intelligent systems leads us to the conclusion that these systems can learn new ways to extract energy and pass information on to descendants, also by activating and deactivating genes. It's certainly not a new idea, and was also upheld by Darwin, who "admits use and disuse as an

---

[6]The paradox is similar to the definition of pornography in US Supreme Court Judge Potter Stewart's ruling: "I know it when I see it" (Jacobellis v. Ohio 1964).

important evolutionary mechanism" not once, but as many as 12 times (see Ernst Mayr's introduction to the *Origin of Species* quoted in Noble, 2010).

On the causality of variations, in the fifth chapter of the *Origin of Species,* Darwin writes:

> I have hitherto sometimes spoken as if the variations – so common and multiform in organic beings under domestication, and in a lesser degree in those in a state of nature – had been due to chance. This, of course, is a wholly incorrect expression, but it serves to acknowledge plainly our ignorance of the cause of each particular variation ... The greater variability of species having wide ranges than of those with restricted ranges, lead to the conclusion that variability is generally related to the conditions of life to which each species has been exposed during several successive generations.

Similarly, in the case of language, not only is there no need to wait for casual variations to make *Homo sapiens* a talking ape "by chance", once again it would appear we must accept a certain level of Lamarckian evolution.

"[Lamarckism] is not so obviously false as is sometimes made out", writes Maynard Smith (1998). "A statement that is all the more significant from being made by someone working entirely within the Modern Synthesis [neo-Darwinist] framework. His qualification on this statement in 1998 was that he couldn't see what the mechanism(s) might be. We can now do so thanks to some ingenious experimental research in recent years" (Noble 2015), thanks to the discovery that "epigenetics can provide a new framework for the search of aetiological factors in complex traits" (Petronis 2010).

## The Anatomy of Language

*Homo sapiens'* organism went through many changes so we could talk the way we do today. In the anatomically modern *Homo,* Daniel Lieberman (1998), (2014) associates the growth of the frontal brain lobes and the hypertrophy of the neocortex located immediately behind the eyebrows, as a possible cause of our faces becoming flatter. While the first feature is a sign of frontal lobes with a greater capability for symbolic reasoning, the second makes room for a vocal tract divided into two parts, a horizontal and a vertical part of the same length – a feature necessary to articulate certain sounds (D. Lieberman 1998).

No other species of the genus *Homo* shares these two features with *Homo sapiens,* so it's highly probable that none of our forefathers could articulate sounds the way we do.

This, however, doesn't mean we were the first to to communicate using some language. As mentioned above some wild gorillas have autonomously developed a form of sign language. The genus *Pan* (bonobo and chimpanzees) uses vocalization to transmit information on available food or the quality of the same (Taglialatela et al. 2003), (De Waal 1988), (White et al. 2015).

Although gorillas and *Pan* have also been going through a process of evolution since our evolutionary lines split, as today their encephalization quotient is lower than that of an *Australopithecus* three million years ago. It would therefore be plausible to believe that similar abilities have been in the curriculum of Hominidae for just as long.

This leads many paleoanthropologists to exclude not only the cognitive revolution of *Homo sapiens* 50,000 years ago, but also the one two million years ago, which according to Louis Leakey led to the emergence of the genus *Homo* (Leakey et al. 1964): "there is no clear evidence of the quantum leap in intelligence and social complexity that Louis Leakey assumed when he first encountered Homo habilis" (Christian 2011).

There wasn't a group of *sapiens* which, by chance or out of necessity, suddenly found it had a "language gene" –a gene of such a reproductive success that it started a new human race.[7]

The cognitive revolutions of Homo sapiens was, in reality, just an evolution, as in the sorite. From some very basic communication capabilities (expressions, attitude, smell) that cannot be defined as language, to today hyper-communication era.

There is another information revolution which makes a good example, because it has not taken millions of years but hundreds: the "IT revolution".

The IT revolution has been less sudden than we first thought. We might start from Blaise Pascal's seventeenth century calculator to the tabulators used to take censuses in late nineteenth-century America, electromechanical calculators, silicon chips, printed circuit boards, and integrated circuits. But none of these technologies resulted in an immediate quantum leap. The first calculators that used semiconductors to compute their results were less powerful than electromechanical calculators, and when the evolution of these could go no further, silicon took over and growth continued. The amount of information we can process with machines has been increasing, in an exponential but continuous way, year after year, since the microchip, or since Pascal calculator, or we could say since the abacus or writing were invented, depending on what you want to consider the origin of the process. In the end, this is "computer science", so we should analyse the emergence of computing, which surely goes back to the Sumeri mathematicians, and so on.

If we take a look at our own built-in computer, the brain, the story isn't much different: "Our data suggest that the evolution of modern human brain shape was characterized by a directional, gradual change" (Neubauer et al. 2018).

In order to understand how our organism allowed language to emerge, we should remember that language itself is an emergent property (Everett 2017). Apparently, everything that concerns our brain is an emergent property, as our neurons are not good for much except communicating with each other.

In the book that introduced modern artificial neural networks, Frank Rosenblatt (1961) writes: "Individual elements, or cells, of a nerve network have never been

---

[7] If we're to take recent history as an example, human society seems more inclined to reward in reproductive terms violent political leaders rather than peaceful geniuses: Einstein, Dirac, Fermi and co. didn't start a new subspecies of *Homo genialis*, while Genghis Khan was such a reproductive success that today his genes can be found in approximately 8% of the males in a region stretching from Northeast China to Uzbekistan", (Zerjal et al. 2003).

demonstrated to possess any specifically psychological functions, such as 'memory', 'awareness', or 'intelligence'. Such properties, therefore, presumably reside in the organization and functioning of the network as a whole, rather than in its elementary parts".

What goes for memory, awareness and intelligence can be said of language too: it's an emergent property of the brain, not the product of a gene that appeared by chance. Excluding divine intervention or pure luck, as we did for the origin of life (Chap. 2), the origin of language may also have been through autocatalysis mechanisms, with neurons and synapses instead of water and an electrostatic field.

In exactly the same way as biological cells emerged from some proteins being able to act as a catalyst for the folding of other proteins, language probably derived from the brain's ability to interpret what occurs in the environment, in a more and more sophisticated way

According to Everett's hypothesis (2017), inspired in turn by Charles S. Peirce's semiotic progression, language initially evolved from *indexes*, in other words indications that something is happening. Smoke indicates there's a fire, tracks that an animal passed that way.

Next steps are *icons*, like the stone an *Australopithecus africanus* was probably carrying around three million years ago – the Makapansgat pebble. Natural elements worked the stone to make it look like a face, and for this reason, the hypothesis is that it was picked up and kept as a kind of amulet or bauble by the Australopithecus.

Then there were *signs* – intentional and arbitrary – and so on, gradually, up to the development of modern language.

What's important to emphasise here in terms of hypotheses like Everett's – or Szathmáry's (2001) *language amoeba hypothesis* – is that language is an emergent property of the brain and as such (like every emergent property) didn't appear suddenly, out of the blue, but through a *bootstrapping*[8] or autocatalysis process. This occurred, thanks to the ability to note, and exploit, minor coincidences: smoke-fire, fire-hot, hot-food smells good,[9] an ability that resulted in the emergence of new cognitive tools.

As mentioned above, this isn't much different from the development of networks of neurons, created to detect the presence of organic material. Hundreds of millions of years ago neurons, which until then had been used to detect glutamate in the

---

[8] One fundamental concept in computer science (although introduced in physics) is bootstrapping. Bootstrapping, in reality, is just another name for autocatalysis, as described in Chapter three with reference to the emergence of cells, but it's also commonly used to explain the growth of systems based on information. This comes from the absurd idea that you can pull yourself up by your bootstraps. On the basis of classic mechanical principles, bootstrapping is impossible, because it creates energy. But, in the case of information and acquiring energy, the process is natural: even a minimum amount of information lets you acquire a little energy, which lets you store a bit more information, and so on.

[9] Without digressing on the use of fire, Wrangham (2009) reports that chimpanzees feed on seeds toasted by natural fires.

environment, a sign of the presence of material that could be used as food, started producing and exchanging glutamate as a means of communication.

The brain took the same path, and *Homo's* brain more than any other. The *problem-solving* capability of the human brain lets it recognise indexes that most other animals miss, just like the neurons that were the only ones that could detect glutamate. *Homo* started with the ability to distinguish indexes, then used this ability to communicate, eventually creating cognitive networks, societies, the equivalent of neural networks of brains.

In time, the ability to recognise indexes was refined: the brain started recognising an albeit casual representation of something that really existed. If a pebble with two holes and its chipped mouth reminds the brain of a face, it will probably do the same for my mates' brain. The ability to turn an icon into a sign is, in a certain sense, the ability to turn the function into an object – to create communication tools.

Indexes and icons in turn therefore act as catalysts for the emergence of complex structures, like signs. This is the *bootstrapping* process. There is no need then to imagine a cognitive revolution that coincides with the birth of language. The use of even an extremely crude form of language lets human societies store and process much more information than other mammals.

The autocatalysis in this case is that, also a primitive form of language – like that used by gorillas – produces a more organised community, more advanced in cognitive terms, and therefore better able to obtain food than others. And the more the communicative society is successful, the more the individual is forced to become more communicative.

The reason why is important that organisms can change their phenotypes in response to environmental stimuli throughout the book, is that this mechanism is essential to explaining the emergence of language: "the origin of human language required genetic changes in the mechanism of epigenesis in large parts of the brain" (Szathmáry 2001).

In the case of language, environmental stimuli are social stimuli. Autocatalysis is the ability of society to force the individual to use instruments for communication – and therefore, if necessary, epigenetic modification.

## Agriculture and Cognitive Social Networks

Another myth that's been busted in recent years, along with that of the cognitive revolution, is the myth of "the invention of agriculture". The idea that there was a moment in which some *Homo sapiens* understood they could grow a plant by planting a seed. The cognitive revolution suddenly made *Homo sapiens* special, but it was the agricultural revolution that decreed the demographic success of the populations that "invented it", the Fertile Crescent first followed by Europe and Asia.

But as Marvin Harris (1978) pointed out 40 years ago, agriculture was probably more the product of necessity than a stroke of genius.

Our forefathers had already colonised a considerable part of the planet's land mass thanks to the social and technological success of *Homo erectus*. The same thing happened with *sapiens*, who developed the abilities of *erectus*. But why so widespread? Because a *Homo* population hunting and gathering needs a notable extension of land to avoid exhausting natural and renewable resources.

The need changes from region to region, but if we consider, as Lieberman did (2014), that a tribe of 20–30 people needs at least 250 square km of land, this translates into an extremely low sustainable population density. To give you an idea of what this would mean, a territory as vast as France for example, even without considering the presence of mountains and less productive areas, could sustain a population of no more than 70,000 – a thousandth of what it does today.

This explains why *Homo's* reproductive success pushed the species to every corner of the planet. If, as surmised by Lieberman (2014), an *erectus* woman on average gave birth to five children, half of which would survive, the growth rate of the population is 0.4% per year.

This doesn't sound like a lot, but it is exponential growth: at this rate, the population doubles every 175 years. A tribe that initially occupied 250 square km, in 4000 years would have grown to occupy an inhabitable surface area of 150 million square km – almost all the dry land that could be used by man. On the basis of a more conservative estimate of our population growth capabilities, Lieberman (2014) estimates that it took *Homo erectus* 100,000 years to spread from Ethiopia to Georgia.

In practice, the advantage with hunting and gathering is that, as it's sustainable, it requires less effort than farming – the first requires no work to transform the environment, the second on the other hand does. But in the case of hunting and gathering, the number of people in the population must remain below sustainable limits. Harris (1978) explains some hunter-gatherer behaviour – female infanticide and the martial tendencies of males – by the need to remain within demographical limits.

Once the *Homo* system had occupied all the available space – the planet – it was forced to stop expanding and had to find a balance. On the one hand there would always have had to be a sufficient number of individuals for inter-tribal and intra-tribal collaboration (the first to renew the gene pool). But, on the other hand, as soon as collaboration made excessive population growth possible, this had to be stopped, through abortion and feuds for example.

Paleontological studies suggest that we have been fighting wars, or little wars (guerrilla) for at least 10–50,000 years (Lahr et al. 2016). But it has probably always been so: aggressiveness must emerge when renewable resources are scarce. When the environment can no longer sustain the *Homo* population, it tends to limit itself to avoid shortages that could lead to severe consequences such as famine.

Agriculture can be considered an alternative to population control. If we modify the environment, it can sustain a larger population.

The "discovery" related to the introduction of agriculture isn't botanical, but once again, social, anthropological. It's difficult to imagine that gatherers 20,000 years ago didn't know that plants grew from seeds (Harris 1978), quite the opposite: although there isn't just one single category of "hunter-gatherers" (Testart et al. 1982), the term "gatherer" is typically used to refer to populations with a more sophisticated botanical culture than that of sedentary populations committed to agriculture (Schultes, Raffauf 1990).

It was therefore necessary to start using agriculture when the fragile balance that was reached when all the available space had been occupied was upset. There may have been many causes of this upset, although climate change has recently been accepted as the principal cause (Richerson et al. 2001). As Cohen (1977) explains: "an imbalance between a population, its choice of foods, and its work standards which force the population either to change its eating habits or to work harder, or which if no adjustment is made can lead to the exhaustion of certain resources."

Agriculture made it possible for man to exploit his intellectual and social skills even more: "crop cultivation fosters association, a desirable goal for our sociable species. At the same time, farming promotes individual ownership and accumulation of material possessions; it makes it easier to have larger families." (Smil 2008).

In practice, agriculture acts as a social thickening agent, as well as being the most reliable method for obtaining food. As mentioned above, the maximum intrinsic rate of increase of *Homo sapiens* was only made sustainable thanks to the unique collaboration capabilities of the species: the more individuals manage to collaborate, the more our species is a success. Agriculture, which requires consistent, coherent communities, means the cognitive abilities of the *community*, of the network of *Homo sapiens*, can explode. Larger and more organised communities encourage the evolution of language, for example with the introduction of the mathematical language, and of communication and information storage, for example writing (Rubin 1995).

"Once food production had thus begun, the autocatalytic nature of the many changes accompanying domestication (for example, more food stimulating population growth that required still more food) made the transition rapid" (Diamond 2002).

Unfortunately the emergence of agriculture, despite the fact that it increased the amount of available energy, didn't make humans abandon their warlike tendencies, quite the contrary. As various tales tell, from Cain and Abel to Romulus and Remus, "[farming] facilitates warfare" (Smil 2008). With a relatively inexhaustible supply of food (when compared to hunter-gatherer populations), the better-organised, more technologically advanced populations were able to expand at the expense of the weaker ones, with relative ease.

Hundreds of books have been written on Alexander the Great's skills in terms of military strategy, but – in terms of information processing – the invention of torsion-spring catapults by the Macedonian army would appear to be just as important as the visionary capabilities of their commander. Without his unbridled ambition Alexander couldn't have conquered an empire, but in a war of sieges, the Macedonian catapults – war machines that could fire projectiles weighing dozens of kilograms at targets hundreds of meters away – were probably just as decisive a factor (Ferrill 2018).

Agriculture made wars of conquest a profitable endeavour. The accumulation of energy, from cereals, means immediate payback to cover the cost of war. Furthermore, rigid social organisation into classes introduced by farming communities –

impossible for hunter-gatherers – favoured the absorption of conquered populations into the new empire: the ruling class might be sacked of its riches, but it was still the head of the organisation. The army was put to death or incorporated into the ranks. The productive classes – peasants – were in the same position as the ass in Aesop's "The Ass and the Old Peasant", with the peasant begging the ass to fly with him as fast it could or else they would be captured by the enemy: "Do you think they'll make me carry heavier loads?" "Oh, well, then," said the Ass, "I don't mind if they do take me, for I shan't be any worse off." (Aesop 1994).

In practice, with the advent of agriculture, war was transformed from a means of population control to an instrument of conquest: great states absorbed weaker ones, to create the first great empires.

## Empires and Networks

The autocatalytic nature of the domestication of plants and animals can actually go beyond "more food stimulating population growth that required still more food", as Jared Diamond wrote (2002). Agriculture acted as feedback mechanism allowing *more people together, who communicate more, who increase their technological expertise, which allow even more people to live together, who communicate more…*

We have seen how the evolution of language into a more and more sophisticated instrument makes it possible for more and more intelligent societies to emerge: able to store information, self-sufficient and self-organising.

When we set up communities of thousands of individuals, a new form of language appeared, mathematics. The power of the mathematical language lies in the use of an extremely powerful syntax that can be used to process a previously unthinkable amount of information.

We've seen how complex the "simple" symbol $\pi$ is. If mathematicians have been obsessed about it for millennia, is because the importance of that number is second only to 0 and 1. The effectiveness of mathematics in natural sciences – in other words in the ability to model and therefore predict our environment – is intimately related to $\pi$, as summed up by Eugene Wigner's tale (1960):

> There is a story about two friends, who were classmates in high school, talking about their jobs. One of them became a statistician and was working on population trends. He showed a reprint to his former classmate. The reprint started, as usual, with the Gaussian distribution and the statistician explained to his former classmate the meaning of the symbols for the actual population, for the average population, and so on. His classmate was a bit incredulous and was not quite sure whether the statistician was pulling his leg. "How can you know that?" was his query. "And what is this symbol here?" "Oh," said the statistician, "this is pi." "What is that?" "The ratio of the circumference of the circle to its diameter." "Well, now you are pushing your joke too far," said the classmate, "surely the population has nothing to do with the circumference of the circle."

But the truth is that the situation described by Wigner appears to be optimistic: many people find it hard to understand that the ratio of the circumference of the circle to its diameter is constant, and will be flummoxed by the fact that said ratio is an irrational (and transcendental) number.

Really there's no need to resort to $\pi$ to have an idea of how unnatural it is for *Homo sapiens* to use mathematical language. Many *Homo sapiens* find it hard to understand that 3/5 is more than half. It's absolutely normal: the human brain is intrinsically non-mathematical – more *communicans* than *sapiens*. "A bat and a ball cost $1.10 in total. The bat costs $1.00 more than the ball. How much does the ball cost?" Everyone, at least for a few seconds, will come up with the answer "10 cents".

We're cooperative primates, with an innate sense of justice, not of mathematics. A chimpanzee, like a person, is happy to work for an apple until it sees its neighbour receives a banana for the same work (De Waal 1996). The ultimatum game[10] experiment shows our sense of justice is innate, and independent of our culture.

"The moral law in us" is, in other words, a common trait. But gazing at "the stars in the sky" and imagining their origin, as did Immanuel Kant, is less so.

The *social* success of mathematics is therefore not due to the fact that, like the spoken word, it's used by the entire population: it's due to the fact that it's the best tool at our disposal for managing public affairs:

> Most pristine state structures depended on organized violence, on religious institutions, etc., and mathematics did not enter. At least one major exception to this rule can be found, however: the earliest "proto-literate" state formation in Mesopotamia of the late fourth millennium, intimately connected to a system of accounting that seems to have guaranteed an apparent continuation of pre-state "just redistribution". (Høyrup 2009)

Not by chance, the word "statistik" was coined around the middle of the eighteenth century to describe the analysis of state data (Achenwall 1748). Mathematics, since the first Sumerian empires, has been society's instrument of management: it first appeared with agriculture 10,000 years ago (Høyrup 2009) and was used to organise society in a relatively fair way.

The passion felt by some *Homo sapiens* for mathematics doesn't, in most cases, derive from the fact that they know it will be useful: for some *Homo sapiens*, mathematics, and logical-deductive reasoning in general, is merely a source of pleasure. When the Sumerians were founding the first civilization, mathematicians gathered data on the position of the planets knowing that in some cases it would take generations to paint a complete picture (Tabak 2004). They worked in the hope and with the conviction that humanity – or in any case the part of the same that dedicated its time to quantitative knowledge – would one day be able to comprehend more complex natural mechanisms than those they could strive to understand with the data acquired during their lifetime.

As a form of language, mathematics, or science in general, is also an emerging property of a community. Something that will survive the individuals.

---

[10] One player, the proposer, is endowed with a sum of money. The proposer is tasked with splitting it with another player, the responder. Once the proposer communicates their decision, the responder may accept it or reject it. If the responder accepts, the money is split per the proposal; if the responder rejects, both players receive nothing. Both players know in advance the consequences of the responder accepting or rejecting the offer. (https://en.wikipedia.org/wiki/Ultimatum_game)

## The Expensive-Class Hypothesis

In the Aztec's "cannibal kingdom" described by Harris (1978), the aristocracy – a combination of the military, priests and mathematicians – had no scruples about keeping most of the population just above subsistence level. Then they would use them as meat to be butchered.

The situation in the Fertile Crescent was completely different. Sumerian slaves were mostly foreigners (Mendelsohn 1946), and there are no traces of ritual cannibalism.

The difference between the two societies wasn't genetic, obviously, but environmental. "Mesoamerica was left at the end of the ice age in a more depleted condition, as far as animal resources are concerned, than any other region" (Harris 1978). The Fertile Crescent, on the other hand, as emphasised by Diamond (2002) "yielded what are still the world's most valuable domestic plant and animal species."

In practice, the Sumerians probably found themselves with a group of priest-mathematicians who, thanks to the presence of resources that were easily exploitable through technological developments, soon turned into a technocracy that made it possible for the society to quickly consolidate itself and expand.

The Aztec aristocracy, on the other hand, found themselves in a situation in which the only way to survive was to become a parasite, considering society as the environment from which they would extract the energy necessary for survival: peasants were considered both workforce and livestock.

We can draw an analogy between these two cases and the brain in Isler and Van Schaik's analysis (2009). The information processing centre – brain or government – is a liability when energy sources are scarce, and in these cases behaves like a parasite – in other words it continues to exist at the expense of the other organs/ classes, part of the same organism/society.

Information processing centres are not only more intelligent, they need more energy too. When energy can be extracted from outside the system, there's no reason to "eat your neighbours". But in times of hardship, the first survival strategy is to exploit productive sub-systems.

The increase of inequality in networks is a phenomenon that's been subject to a great deal of study. As pointed out by Dorogovtsev and Mendes (2003), in a society in which wellbeing decreases, the level of inequality increases.[11] Seen as intelligent systems, government or the brain not only need more energy, they're also in the position to subjugate the rest of the organism to obtain it.

Although new technologies, from the brain to mathematics and computers, are developed in periods of opulence – like the electronic calculators developed in post-World War II America – the same technologies will be used to the disadvantage

---

[11] Dorogovtsev and Mendes measured the level of inequality, and rightly so, not with indexes like the Gini coefficient, but by measuring the exponent of the power-law which describes the distribution of wealth (see Appendix 1). The greater the absolute value of the index, the greater the inequality.

of the common good and to the advantage of the governing minority in times of need.

We'll take a look at how *Homo sapiens* evolved after the invention of language, and how human societies evolved through the evolution of communications' instruments in the next chapter: in practice, whether cells or societies, systems that can effectively process information can exist as a parasite in another system (the Aztec ruling class or Dyson's RNA) or as organisational centres for the system as a whole.

In the following chapter we'll see how something similar is happening in modern *Homo sapiens*' societies with the emergence of today's IT giants.

# Chapter 5
# The Human Meta-Organism

The aggregation of cells into complex organisms is considered a new form of life, although it's really just the result of the aggregation of other lifeforms. This makes sense though: when cells reached their cognitive limit, by connecting, intelligence continued to evolve. The cognitive capabilities of complex organisms have little to do with the ones of a single cell.

We've also seen that the above development occurs seamlessly: we go from the distributed collaboration of bacteria up to complex organisms.

Bacterial colonies aren't really organisms: they don't appear to have any self-awareness, they cannot clearly distinguish between "themselves" and the external environment. An organism, on the contrary, knows which elements are part of the system (and therefore collaborate in the cognitive processes, and must receive energy) and which aren't (and can therefore be sources of energy).

Networks of complex organisms –societies– however, are more similar to bacterial colonies than to a new organism. The only possible exception, colonies of insects, are somewhere between the two: they are formed by extremely simple complex organisms, like ants, whose network creates a superorganism, like the ant colony. But the colony's cognitive abilities are well below the ones of many complex organisms.

In this sense we say that the appearance of complex organisms is a revolution in the evolution of life. Complex organisms went on developing cognitive abilities that no single cell can have. Networks of complex organisms, like ants or social mammals and birds, on the contrary *did not* show such a leap in intelligence.

The aim of this chapter is to show how the introduction of communication through electric signals by *Homo sapiens* pushed the cognitive abilities of our species: not the ones of the individual, but the ones of human society as a whole, which, with the introduction of Internet, is starting to look more and more like a complex organism made of complex organisms.

This, in geological terms, is indeed a historic moment: the emergence of the first meta-organism of complex organisms, the human meta-organism. A new milestone in the evolution of life.

© The Author(s) 2020
M. Alemi, *The Amazing Journey of Reason*, SpringerBriefs in Computer Science, https://doi.org/10.1007/978-3-030-25962-4_5

## The Evolution of Communication in *Homo sapiens*

Hippocrates of Kos' quote "Ὁ βίος βραχύς, ἡ δὲ τέχνη μακρή" was translated into Latin as *Vita brevis, ars longa*. The word τέχνη (téchne), from which the term *technology* comes, derives from the Proto-Indo-European root *teks-, initially "to construct putting together" and then "to weave", from which the Italian word *tela* (cloth) derives. The origin of the Latin *ars* is similar, and derives from the Proto-Indo-European $h_a er$-, "prepare, put together" (Mallory, Adams 1997). Taking liberties with the quote one might paraphrase it as "the individual dies young, the network lives long".

Hippocrates' quote is reminiscent of Sumerian mathematicians: a life is not enough to collect all data needed to build a model. The solution adopted by the Sumerians was to introduce a form of language, mathematics, and a communication channel, writing. This, as described in the previous chapter, increased their ability to store and process information as a network of individuals. In short, it created an intelligent system that survived longer than the single components of the same, and even longer than the system itself: while the Sumerians have disappeared, their math lives on.

The invention of mathematics and writing can be compared to the introduction of synapses by neurons. Spoken language in fact, on its own, has some clear limitations, as it did for early neurons communication through the diffusion of chemicals. It's difficult to imagine *Homo sapiens* communicating *verbally* with 7000 or 20 million people, respectively the average and highest number of neuron synapses in our brain.

*Homo sapiens*, as Dunbar (1992) proved in his study on the relationship between the cortex and the size of communities, can't connect to more than about a hundred other individuals. If mammals, and *Homo* first and foremost, managed to use the network effect for leverage, despite the small size of the same, this is also and especially thanks to the impressive cognitive abilities of individuals.

As mentioned above, the brain was originally a processor of information, and evolved into a communicator, in the same way as the neuron was originally a detector of edible material that evolved into a communicator cell.

In practice, social networks of mammals are mainly a multiplier of individual intelligence. They are not a new entity, a real meta-organism, from the word μετά: "after", "beyond".

## More Communicans than Sapiens

In an information-energy context, it's this thrust towards more intelligent cognitive networks, systems that can process more information in order to obtain more energy from the environment, that drove the *Homo* brain to become constantly more powerful for millions of years.

But as mentioned in the previous chapter, it's a substantial investment for a mammal to keep its brain functioning. Increasing cognitive ability more than *Homo sapiens* has done would be risky and impractical in energy management terms.

According to Hofman (2014), *Homo sapiens* brain could reach its maximum processing power, approximately 50% more than it has today, by increasing its volume by 130%. If so, it would weigh almost 3.5 kg compared to the 1.5 kg it weighs today.

Leaving aside the question as to whether our organism could physiologically sustain a greater brain mass, it's obvious that 100,000 years ago a brain that was a bit more powerful but in proportion required a lot more energy, would have been nothing more than a risky liability.

So, it was a much better idea to develop communication abilities, creating the necessary specific instruments: doing the same thing as neurons, that created synapses and dendrites in order to communicate more effectively.

Put simply, the first instrument of communication to emerge after language, was writing. The written word, at least in the Fertile Crescent, was probably developed to keep accounts in the taxation system: the oldest example of writing is a Sumerian income statement, not a poem (Lerner 2009).

Writing was invented to boost the cognitive abilities of some brains, those of the Sumerian civil servants-mathematicians, so they could create small cognitive networks.

As communication developed through writing a change began: from storing and processing information as an individual to doing so as a network. This became evident to some *Homo sapiens* already thousands of years ago. In Plato's Phaedrus, Socrates exclaimed:

> …this discovery of yours [writing] will create forgetfulness in the learners' souls, because they will not use their memories; they will trust the external written characters and not remember themselves. The specific which you have discovered is an aid not to memory, but to reminiscence, and you give your disciples not truth, but the semblance of truth; they will be hearers of many things and will have learned nothing; they will appear to be omniscient and will generally know nothing; they will be tiresome company, having the show of wisdom without the reality.

It's a concept that's repeated over the centuries, every time a new instrument of communication emerges. In 1492 the German Renaissance polymath Johannes Trithemius published *De laude scriptorum manualium* (In praise of scribes). Trithemius considers manual writing to be a form of higher learning:

> [Writers,] while they are writing on good subjects, is by the very act of writing introduced in a certain measure into the knowledge of the mysteries and greatly illuminated in his innermost soul.

But as Ziolkowski (2011) mentions, "Trithemius himself was no foe of printed books", and ironically, the treatise has survived till today only as a printed work. The same is true for the words of Socrates, which we can read only because his pupil Plato transcribed them.

If we continue to follow evolution in instruments of communication, the telegraph first and then the telephone were received in the same way: they made it possible for a global society to emerge, but were also attacked.[1]

At the end of the nineteenth century the American writer C. Harris wrote:

> At present our most dangerous pet is electricity – in the telegraph, the street lamp and the telephone ... The telephone is the most dangerous of all because it enters into every dwelling. Its interminable network of wires is a *perpetual menace to life* and property. In its best performance it is only a convenience. It was never a necessity (Harris 1889).

Similar stands were taken against the radio, television and, obviously, against communication through interconnected networks managed by electronic calculators, the Internet, and the subsequent development of graphical interface for non-professional use, the World Wide Web.

Perhaps, the first *Homo* who used a form of spoken language to settle controversies in a group was criticised by the older generation too. How much less "humanity" was there in communicating verbally, at a distance, compared to the physical contact of *grooming*, of reciprocally cleaning each other's fur? We'll never know, because the older Hominidae probably conveyed their disapproval with a few grunts of disgust, and that was the end of that.

Socrates, Trithemius and those who today complain about the externalisation of our memory from the brain "to Google" have a point though. Language, writing, print, have resulted in societies able to extract more energy from the environment, to such an extent that we can now sustain a population of several billion *Homo*, not necessarily any wiser.

As mentioned in Chap. 3, quoting Rousseau's *On the social contract*, the more society acquires cognitive abilities, the more the individual becomes insignificant in relation to the rest.

The more *communicans Homo* becomes, the less he is worthy of the name *sapiens*.


## Communications Technologies and Topologies

On the basis of the brief overview of communications technologies in the previous section one might think there had been a constant improvement in means of communication, as communications channels have constantly increased. But it's not that simple.

Despite the fact that today there's more focus on "bandwidth" in economic development policies, it's also, if not above all, the topology of the network that determines the amount of information a network can process.

---

[1] Thanks to David Malki for his blog http://wondermark.com where all the examples quoted were found.

In this sense, instruments of communication invented by humans, regardless of the capacity of the channel, ie the amount of information/syntaxes that can flow from one element to the next, can't be considered a constant improvement in information processing capacity for society.

The *C. elegans* neural network analysed in Chap. 3 is nothing special in terms of amplitude of the communication bandwidth between neurons, or number of nodes. But as we've seen, this type of network makes the processing of sensory input surprisingly accurate.

The first societies based on spoken language in a certain sense represent the equivalent of the *elegans* proto-brain: each person can exchange signals with any other member of the group.

The same thing can't be said of the written word. On the one hand, writing is in itself a way to store information, with the obvious advantage that something written today can be read centuries later, as it happened to Socrates. It can therefore be used to create information networks that develop in time, and not just in space.

But on the other hand writing comes with some distinct disadvantages, including the externalisation of memory. It is for example quite an inflexible form of memory, a bit like DNA. Languages that aren't officially used in a written form are more plastic than written ones (Hollenstein and Aepli 2014). Languages without a written tradition are not less expressive than official languages, quite the contrary. Consider the importance of dialect poetry in countries that have an official language such as Italy, or the acknowledgement that "the highly verbal" African American Vernacular English "is famous in the annals of anthropology for the value placed on linguistic virtuosity" (Pinker 2003).

The fact that in Switzerland 80% of the population uses a non-written language as their first language (this is the percentage of people who speak Swiss German and Italian dialects) maybe due to the "early and long-lasting interest in pragmatism" in the country (Tröhler 2005).

The Latin saying *verba volant, scripta manent* can be interpreted in two ways: written words are of more certain interpretation than spoken ones, or spoken words let you fly away, written words keep your feet on the ground.[2]

Regardless of the speculations concerning the connections between the nature of a people and the use of writing, what's certain is that writing started a transformation of human social network. Whereas before, one person could have bidirectional contacts with about one hundred others, now one single written work created a *one-way* channel between writer and readers. Writing is a "one-to-many" communication system. Intellectual currents do form around written works, but societies cannot participate as a whole. First writing, and then printing, in the best case scenario create small connected sub-networks (the intellectuals) whose ideas propagate until they eventually touch the society.

The situation changed radically with the invention of the telegraph, for two reasons. The first is that, for the first time in the history of complex organisms, these

---

[2] I was told the second interpretation by my late father, who loved Latin and Ancient Greek.

organisms managed to communicate with each other at speeds close to that reached by the nervous system. The second is that the telegraph made it possible to create a network that was topologically similar to neural networks, in which each individual could connect to all other individuals.

"*Could* connect" because in reality the telegraph is an instrument that's problematic to build and keep working over long distances, and it's also complicated to use. But it was still an *embryonic* nervous system, as was obvious to Carl Friedrich Gauss almost two centuries ago.

Gauss, who as well as being one of the most brilliant and prolific mathematicians and physicists of all time, never turned down an opportunity to experiment, in 1833 wrote to the astronomer Heinrich Wilhelm Matthias Olbers: "I don't remember any mention to you about an astonishing piece of mechanism we have devised." (Dunnington et al. 2004).

That piece of mechanism was the telegraph, which Gauss and his young colleague Wilhelm Eduard Weber had invented, built and installed so they could quickly communicate between the observatory and the institute of physics. Gauss and Weber published their results in German in the *Göttingische gelehrte anzeigen* (Dunnington et al. 2004). Despite the initial interest shown by politicians (it was presented to the Duke of Cambridge), the invention wasn't a success, also because after the two inventors proved the feasibility of the project they didn't have time to dedicate to its industrial development.

But the importance of the discovery was clear to Gauss from the start. In 1835, the scientist wrote to his ex-student Heinrich Christian Schumacher: "The telegraph has important applications, to the advantage of society and ... exciting the wonder of the multitude". In the same letter he also mentioned the estimated investment necessary to lay the required wires around the world: 100 million thalers.[3] He concluded by mentioning that he had found it easy to teach his daughter to use the instrument (Dunnington et al. 2004).

At the time, no one shared Gauss's enthusiasm, and the telegraph was reinvented and developed by others. It's easy to understand Gauss's contemporaries, as it must have been hard to imagine how two needles miles away that move in synchrony with each other could one day "excite the wonder of the multitude".

What Gauss was rightly enthusiastic about was the possibility of communicating *instantaneously* between one place and another on the planet. To create, to all effects and purposes, a network of electrical synapses in which the people are the neurons. It took humanity almost 200 years to evolve the telegraph into something that could arouse the interest of the masses. This was because the necessary investment was huge, and because the technology wasn't easily scalable. A wire one mile long is one thing, a network of wires around the planet, a "world wide web", a completely different concept.[4]

---

[3] Considering the observatory's budget was 150 thalers (which Gauss was known to complain about), the investment would be about 100 billion dollars today, a more than reasonable estimate.

[4] The telegraph in topological terms can be considered to be a network that's similar to that of the brain, but in terms of physical connections it's different, and this made development difficult: if a

But all things considered, the result is the same: each element can communicate in a bidirectional way, almost instantaneously, with a very high number of other elements, which is what matters in terms of processing information in a network.

This means the telegraph and its spin-offs are perfect to create the "next level" of intelligent systems: they can transform a social network into a network that's topologically similar to a network of neurons, with similar cognitive abilities.

Systems such as radio and television on the other hand, with their one-way star structure (a few transmitters and many receivers) are exactly the opposite. They're used to connect one node to others and not vice versa.

If we consider this in terms of Hebbian learning, radio and television don't let the network learn much: there's one single central neuron, connected to peripheral neurons, that acts in a relatively independent way. As predicted by Hebb, connections will be created between peripheral neurons only because they're in tune with the central neuron, which occurs often, considering its power.

So Hebbian learning, considering the topology of the network, was responsible for the fact that radio was a highly effective instrument of propaganda for early twentieth-century governments. As Joseph Goebbels said soon after the Nazi party came to power in Germany in 1933: "It would not have been possible for us to take power or to use it in the ways we have without the radio…" (Adena et al. 2015).

## Internet Companies

When we talk about Artificial Intelligence we mostly think of companies like DeepMind, an Alphabet Inc. (former Google Inc.) subsidiary, with neural networks that can learn to play Go or video games, or certain products like chabots and speech-to-text, rather than algorithms that are currently used by Google, Facebook, Amazon and Netflix (the FAANGs excluding Apple, or the FANGs) to make a fortune.

Although the technology is very advanced, algorithms, in terms of mathematical sophistication, are less so. But the ability to extract a great deal of information is notable, and this gives these companies a huge advantage over everyone, governments included.

One of the most obvious examples is Alphabet Inc., which continues to generate a healthy revenue through the Google search engine. Google's ability to find the most influential nodes in a network is based on a brilliant algorithm published by the two founders. This algorithm, called PageRank, calculates the central position of the nodes (Brin and Page 1998) in a very precise way, and could finally be implemented by Google's engineers on a scalable and cheap infrastructure. It really was

---

telegraph can connect to 10,000 other telegraphs it can't have 10,000 outgoing connecting wires, like the synapses of a neuron. Hubs are required to route the communication between two elements: this was done for the telephone by switchboard operators, until a few decades ago, and is now automated using specific computers called routers.

a silver bullet. But it wasn't just the PageRank that made Google one of the most successful companies in the history of finance.

PageRank and its implementation allowed Google to offer an excellent service, fulfilling its *mission*: "To organise the world's information and make it universally accessible and useful".

But what makes Google still today such a successful company is the application of smaller strategies, much less sophisticated than PageRank, used to extract energy (revenue) from the external environment. An environment which, in this case, consists of its users and the companies who want to reach those users.

Today, Google has managed to make the dreams of some CERN researchers of the 1990s come true: to make people pay a fee for each site they browse.[5]

In its early years, Google effectively put its mission of organising information into practice. The compromise was that the service wasn't monetised through payments (the founders' initial idea), but by proposing sponsored links, always pertinent to the user's search and clearly separated from the main results.

People looking for "Lawyers in Paris" would find various websites ranked by relevance, plus others, clearly highlighted as "sponsored links", relegated on the right of the page. Sponsored links were ranked by relevance *and* by the amount of fees paid to Google.

The breakthrough arrived with the analysis of user behaviour. The company noticed that many of the search queries it received weren't keywords, but names of web site. In other words, people who wanted to visit the site of the company ACME, whose web address was http://www.acme.com, didn't type such address in their browser. They would *google* "ACME" and then click on the first link proposed, which infallibly (and easily) was redirecting them to http://www.acme.com.

This is when Google started having the signal "Sponsored links" changed to a timid "Ad", and serving those paid links *on top of the real results*. In this way, ACME, in addition to optimise its content according to what Google wants, has to pay to be on top of the paid links.

ACME must pay, because if it does not pay enough, competitors might appear at the top.

How little importance Google puts on linking the user to the web site they actually wanted to reach is clear in Fig. 5.1. In this case, the first result in a search for "SK Traslochi" is the competitor "traslochi 24". The link to the competitor is labeled "Ann.", which has no meaning whatsoever in Italian, surely not "Sponsored".[6] Users looking for "SK Traslochi", used to trust Google, are easily tempted to click on the competitor's link. That *is* evil, and is far from Google's original mission.

Google has other revenue channels apart from its search engine, all based on analysing user behaviour. For example, thanks to a "free" tool used by websites to

---

[5] At the time, I heard many such complaints from my colleagues, who didn't realise that, if the Web won over all other solutions (does anyone remember Gopher?), it was precisely because it was free of any commercial license, see (Berners-Lee and Fischetti 1999).

[6] Verified 7 April 2019.

**Fig. 5.1** Google proposing a competitor's website as first (sponsored) link

analyse web traffic, called Google Analytics, a huge number of websites[7] send to Google detailed information on who is reading what.

It is interesting to read in the book written by Google's first director of marketing, Douglas Edwards (2011), how the idea to just insert sponsored results was initially considered immoral by the company's employees.

Now, however, Google gives companies paid-for visibility: if you want to appear as the first link when your customer *google* your company's name, you have to pay. If your competitors have a good Search Engine Optimization expert, you might have to pay a lot.

The model described above is in line with the real mission of Google's holding, obvious if we look at the name: *Alpha-bet*, a bet on alpha, the symbol used in finance as a measure of the excess return of an investment in relation to the market benchmark.

Alphabet's *raison d'être* isn't to provide the perfect website, but rather a website that will satisfy the user, making them pay for it, albeit in an indirect way. *There is no such thing as a free lunch*: users think they're not paying, but companies have to pay for the user to find them, and as a consequence must increase their prices to compensate for these additional costs.

With Google Analytics, companies think they don't have to pay for a service. They do have to pay Alphabet for their products and services to be displayed to their target users though, something Google is well aware of. Google, in fact, collects non-aggregated browsing data, in other words it knows exactly which person visited the website, but it only provides Google Analytics users with aggregated data. What's more, Google can cross-reference visit data with search data, to identify the user profile precisely.

Similar strategies are used by Facebook to circulate posts, and by Amazon after it introduced "sponsored products".

---

[7] Google does not provide any figures on the number of websites using its product

## *On Regulating the Private Sector*

As Yoshua Bengio says, when powerful algorithms are used exclusively for company profit, this creates dangerous situations:

> Nowadays they [the big companies] can use AI [Artificial Intelligence] to target their message to people in a much more accurate way, and I think that's kind of scary, especially when it makes people do things that may be against their well-being (Ford 2018).

Bengio isn't exaggerating, although it does not mean that the role of these companies has never been beneficial.

While the public sector was responsible for the impetus behind the creation of the World Wide Web, it was the private sector – companies like Google – that made the invention useable by the masses. Reading Edwards (2011), one sees Google during the first years of existence as a community of *hackers,*[8] whose purpose was effectively to organise global information and, above all, solve the technological problems that made indexing billions of web sites more and more difficult.[9]

This vocation for problem solving, for "making the world a better place", is one of the mantras of all technological start-ups, right up to the day they're listed on the stock exchange. As Cringely (1996) explains, IT companies have to aim to be listed on the stock exchange not to acquire capital (in relation to any other industry, IT is not very capital intensive), but to liquidate workers, who are paid in company stock options. The workers have to be paid in *stock options* because the good old *hackers* were inclined to leave the company after solving the first intellectually stimulating problem they were given, to look for another problem. It's only the mirage of millions of dollars that lets these companies keep the first wave of creative minds, the ones that can solve the most challenging problems, on their payroll for more than a couple of years.

When the company is listed on the stock exchange, the first wave of *hackers* jump ship, and the company organisation is set up. But the price of the shares on the market must always continue to grow, because if it doesn't there'll be an excessive brain drain and, as a consequence, a loss of capital. A vicious circle that would destroy the company.

In short, the company's *mission* becomes: to increase revenue.

Google found itself at the right place at the right moment: it received major funding just before the *new economy* bubble burst in 2001 (Edwards 2011). This let the company scrape together the finest computer scientist and solve problems that had until then been considered unsolvable.

If, as Tim Berners-Lee wrote, "…people say how their lives have been saved because they found out about the disease they had on the Web, and figured out how

---

[8] "Hackers solve problems and build things, and they believe in freedom and voluntary mutual help." Eric Raymond in *How To Become A Hacker*, http://catb.org/~esr/faqs/hacker-howto.html, verified 12 April 2019.

[9] "The Friendship That Made Google Huge" by James Somers, The New Yorker, December 3, 2018

to cure it",[10] the credit goes to Google too. But when Edwards (2011) describes Google's employees being told the company would be listed on the stock exchange, we see kids who have won the lottery rather than people who want to make the world a better place. The result is that, as Berners-Lee said on the 30th anniversary of his proposal for an information management system at CERN in Geneva, "user value is sacrificed, such as ad-based revenue models that commercially reward clickbait and the viral spread of misinformation".[11]

In the west, as well as the above-mentioned technology giants listed on the stock exchange, there are also new entries that have taken advantage of the power of the Internet to solve real problems, as Google did in the 90s.

One example is AirBnb. There are many advantages for the economy, as confirmed by independent studies (Quattrone et al. 2016) and, of course, by AirBnB itself. The services offered by AirBnb in some cases are more efficient than the services offered to tenants by the state. In Italy for instance, unpaid rent amounts to 1.2 billion euros per year, and it's hard to imagine recovering the debt you are owed in less than a year.[12] With the law incapable of guaranteeing fulfilment of a contract, AirBnB is considered the only viable option for renting out a house.

AirBnB is in the same situation that Google was in around the year 2000, or Facebook a few years later, but if left unregulated there's nothing to guarantee it won't represent the same risks as Google and Facebook today.

Facebook, for example, has developed a technology that's ideal for someone who wants to make harmful information go viral. Let's take vaccines for example. Citizens/users find themselves torn between two contrasting sources of information: on the one hand medicine, which wishes to assure them that the probability of infection is minimised; on the other, a user or organisation which, in good faith or not, spreads the word that vaccines are harmful.

A Facebook user, the father or mother of a child, who in their timeline sees a post entitled "Vaccines reduce the probability of infection in children" won't give it much consideration. But "Vaccines cause autism" shocks the user, who hesitates while scrolling the timeline. The Facebook algorithm is not designed to minimise deaths from infection, but user engagement, and slowing down while scrolling means more engagement.

This is enough to display the post against vaccines to other users too and hide those promoting vaccines: the time each user spends on the application, along with the number of users, is one of the most important parameters for investors. The number of deaths of children who weren't vaccinated don't appear in investor relations.

---

[10] Tim Berners-Lee, "Answers for Young People", https://www.w3.org/People/Berners-Lee/Kids.html, checked on 12 April 2019.

[11] Tim Berners-Lee, "30 years on, what's next #ForTheWeb?", https://webfoundation.org/2019/03/web-birthday-30/, verified 12 April 2019.

[12] La Stampa (Italian newspaper), 14 August 2018, tenants in arrears, a portal for recovering unpaid rent https://www.lastampa.it/2018/08/14/economia/inquilini-morosi-arriva-il-portale-per-recuperare-gli-affitti-non-pagati-Sj3au6263cQsApQAKw6VEN/pagina.html, verified 12 April 2019.

Facebook acknowledges that some of its users create fake news for economic gain,[13] but does not acknowledge that Facebook Inc. also benefits from fake news. Facebook makes profit every time the user clicks on a sponsored link, regardless of the content. Posts containing disinformation are the perfect instrument for identifying the ideal user for those who want to sell you something, first and foremost politics: always on the lookout for gullible people.

Let's take Italy, the country that in the last century was the testing ground for the rise and consolidation of fascism. In the *bel paese,* the two political parties that formed the Italian government in May 2018 (the Five Star Movement and Lega Salvini Premier) both promoted anti-vaccination policies while in power (Sole 2018), (Repubblica 2019).

Some Italian Five Star Movement members of parliament publicly uphold the existence of chemtrails and Judeo-Masonic conspiracies, not to mention the fake moon landing and mermaids[14].[15]

Populist parties feed off gullibility. There are no limits to the pre-electoral promises a "flat earth" voter will believe, or to the excuses given, without fail, after the elections.

Once gullible users have been identified, they can be targeted with the most unbelievable messages, from immigration being the cause of the economic crisis to the phantasmagorical profit to be made from exiting the European Union.

Not by chance, the Five Star Movement is the political arm of a communications company. One that understood before others how to use digital channels (Casaleggio 2008). Likewise, the Lega Salvini Premier makes use of a seasoned digital communications' team (Espresso 2018).

The spectrum of disinformation is wide, and obviously Facebook and other platforms are not the only cause of disastrous political decisions. For example, a variety of factors made 17 million British citizens vote in favour of Brexit (Kaufmann 2016). But the 350 million pounds/week that Brexit should have speared to the British economy (Rickard 2016) remains a masterful ruse, or a criminal way to use digital communications' channels, depending on how you look at it.

Facebook is unrivalled in being able to find gullible user segments. The statistical analysis done academic to find out which human profiles voted for Brexit (and therefore also their reasons) are certainly sophisticated, but also based on a ridiculously small sample when compared to the "big data" of the company. Two examples:

---

[13] "We're getting rid of the financial incentives for spammers to create fake news -- much of which is economically motivated." Mark Zuckerberg, Second Quarter 2018 Results Conference Call, https://s21.q4cdn.com/399680738/files/doc_financials/2018/Q2/Earnings-call-prepared-remarks.pdf

[14] "Carlo Sibilia, the Five Star Movement's conspirationist, the new Interior Ministry Undersecretary", La Repubblica, 13 June 2019, checked https://www.repubblica.it/politica/2018/06/13/news/carlo_sibilia_sottosegretario_all_interno-198873517/)

[15] "Secrets and chemtrails: the long list of Five Star conspiracies", Espresso, 26 September 2014, verified 9 April 2019: http://espresso.repubblica.it/palazzo/2014/09/26/news/l-eurodeputato-m5s-a-caccia-di-scie-chimiche-ecco-la-lunghissima-lista-dei-complotti-grillini-1.181876

Kaufmann (2016) reached the following conclusion "primarily values ... motivated voters, not economic inequality" after analysing the results of a survey that interviewed 24,000 people. Swami et al. (2017) reached the conclusion that people who believe in Islamic conspiracies are more likely to vote for Brexit thanks to an analysis of an opinion poll with 303 participants.

Let's compare this figure with Facebook use in the UK at the time of the referendum: 37 million users,[16] with up to 2 hours per day spent using the app.[17]

Technology giants have more data on habits, behaviour and opinions than any other human organisation. In 2017, "only around 43% of households contacted by the British government responded to the LFS [Labour Force Survey]", a survey which is used to prepare important economic statistics in Great Britain.[18]

On the contrary, Facebook and Google know where users are, who they are acquainted with, what they are watching. Google users, through a simple search, tell Google their wishes and problems, things they probably haven't told anyone else, and they didn't have to answer even one survey question to do so.

The problem with big Internet companies is that their organisation and capabilities could almost be considered those of a global brain. Their behaviour however cannot be considered in the same light.

Behind the success of complex organisms' brains there's always a balance of costs-benefits. As mentioned in the previous chapter, the *expensive tissue hypothesis* proposes the idea that the human brain, becoming more and more costly in energy usage terms, made the organism sacrifice part of its essential organs such as the digestive system or the locomotor apparatus. But this didn't create problems for the organisms, quite the contrary.

If we look at the brain as an independent system, the brain always considered itself to be part of the organism, and always identified the environment *outside* the organism as the source of energy it needs to survive. The organism is an organisation of collaborating organs. They brain does not feed itself at the expense of other organs.

Big Internet companies on the other hand, see human society as the environment from which they extract energy. In the best case scenario they can be compared to parasites – foreign organisms that feed off their host. In the worst case scenario they are like tumours – sub-organisms that grow out-of-control, and that in order to maintain their level of low entropy are willing to sacrifice the very life of the organism of which they themselves are part.

---

[16] Forecast of Facebook user numbers in the United Kingdom (UK) from 2015 to 2022, https://www.statista.com/statistics/553538/predicted-number-of-facebook-users-in-the-united-kingdom-uk/, checked 9 April 2019

[17] Average daily usage time of Facebook in the United Kingdom (UK) 2014, by age and gender, checked 9 April 2019, https://www.statista.com/statistics/318939/facebook-daily-usage-time-in-the-uk-by-demographic/

[18] The Economist, 24 May 2018, "Plunging response rates to household surveys worry policymakers", https://www.economist.com/international/2018/05/24/plunging-response-rates-to-household-surveys-worry-policymakers, verified 9 April 2019

This might seem excessive, but the number of deaths of people who haven't been vaccinated could be just the tip of the iceberg. This is the mechanism with which new nationalist, populist or openly fascist movements including Donald Trump in the US, the Five Star Movement and Northern League in Italy, Narendra Modi in India (the first to use WhatsApp,[19] a service owned by Facebook, in politics) and Jair Bolsonaro[20] in Brazil came to power. They all exploited so-called *social networks*.

Trumpeting about making the world a better place, the *big tech* have become similar to their parodies: in the *Silicon Valley* TV series, Gavin Belson, the CEO of the fictitious Hooli Inc., clearly based on Google, says: "I don't want to live in a world where someone else makes the world a better place better than we do."

In conclusion, on the one hand there are companies that provide services, from web indexing to renting homes, machines and hiring labour, in a much more efficient way than states.

On the other hand, there's the problem that the ultimate aim of these companies is to increase revenue, whatever the cost. The aim is not to improve the lives of the people.

Probably and hopefully in the future both the value of Internet companies and the need to regulate their actions will be acknowledged, in exactly the same way as is done for water and electricity[21] today. In practice, this will force these systems to acknowledge their role as part of society.

## The Evolution of Artificial Intelligence

While artificial and natural neural networks have some things in common, in the context of this book it doesn't make much sense to ask oneself if one day artificial neural networks will be more intelligent than *Homo sapiens*.

Artificial intelligence might have been created by *Homo sapiens*, but it is no less natural than its creator, or any mechanism that other forms of life, as intelligent systems, developed to extract energy and feed their cognitive abilities. It is always natural evolution which led to the introduction of the *C. elegans* neural network, and "artificial" neural networks a few million years later.

One thing it does make sense to ask is why Artificial Intelligence systems emerged, why in this form, and what role will they play in the evolution of life on earth. Or rather: how will their role evolve, in consideration of the fact that they already play an essential role in our society.

---

[19] "India, the WhatsApp election", The Financial Times, May 5, 2019.

[20] "How social media exposed the fractures in Brazilian democracy", The Financial Times, September 27, 2018

[21] The Economist, 23 September 2017, "What if large tech firms were regulated like sewage companies?", https://www.economist.com/business/2017/09/23/what-if-large-tech-firms-were-regulated-like-sewage-companies, verified 12 April 2019.

The reason why we can't remain indifferent to artificial neural networks is that, unlike other Artificial Intelligence systems, they are incredibly autonomous systems. It is as if *Homo sapiens* effectively created a sort of brain, and then the brain learnt on its own. Mathematicians create the structure and expose it to the environment. The structure, autonomously, not only learns, it adapts so it can represent the environment.

Considering how the cognitive ability of artificial neural networks can evolve autonomously, it's essential we realise why we reached this point, and which direction we might take.

An important aspect of artificial neural network is that *Homo sapiens*, scientist or not, has little to say on how the system will behave. Today, data is the true mine of information, and no longer the mathematical model making sense of those data. When you have enormous amounts of data, mathematical sophistication becomes less important.

This is something old-school scientists, like physicists, have had difficulty recognising, contrary to computer scientists. In the aptly titled "The Unreasonable Effectiveness of Data", three famous researchers (Alon Halevy, Peter Norvig, and Fernando Pereira (2009), all Google employees) refer to Wigner's book (1960) mentioned in the previous chapter.

The article was written when the academic world, and not only, had accepted a return to neural networks, after the "long winter" of the 1980s and 1990s. Although the article doesn't mention neural networks, it predicts exactly what neural networks would soon be capable of in the near future: thanks to the analysis of a significant amount of data, machines would be able to perform tasks that had been unimaginable until then. The sophistication of the mathematical model that explains the phenomena is less important than being able to predict them.

At the end of the day, that's all intelligent systems have to do. Creating mathematical models is a characteristic of no other complex organism except very few *Homo sapiens*. Most intelligent systems reduce uncertainty using mechanisms similar to neural networks, i.e. without trying to understand why. It's not surprising therefore that in artificial neural networks we've developed something that works, in cognitive terms, in quite a similar way to other intelligent systems, like the *C. elegans*[22] brain. As of today, with big limits though.

The algorithms used today are mostly classifiers. In other words, attempting to maximise a gain function the algorithm put an input in a probability box: label "A" has a certain probability of being true, label "B" another probability and so on.

In a game of chess, the gain function is winning. After having analysed hundreds of millions of games, a neural network can predict that a certain move will increase the chances of winning. But we're still a long way from being able to say machines are *sapiens* though.

---

[22] One of the reasons why it was difficult for artificial neural networks to emerge is because a system that incorporates the programmer's logic (an "expert system") functions immediately, without the need for training. But it's more difficult for them to learn new strategies when the environment changes.

In 1950, Alan Turing (1950) developed what today is called the "Turing test": a machine exhibits human cognitive abilities if a human can interact with it (by chatting for example) without being able to tell they are talking to a machine. As some people say, either the machine is very competent, or the person is not. In any case, it does not seem to be a definitive test for measuring how "human" and algorithm is. And typically, algorithms in Artificial Intelligence are not build to behave as humans. In Artificial Intelligence we look for other capacities, that humans don't possess.

If Ke Jie, the Go champion who lost to AlphaGo (Chao, 2018), had played against a remote machine, he probably wouldn't have known his opponent was not human. But this doesn't change the fact that AlphaGo does not learn or think like a human being: a person learns to play Go in just a few minutes, by simply reading the rules. If Ke had had to teach the machine to play Go, he would have immediately realised there was not a person on the other end of the line. The machine will never be able to think like a *Homo sapiens*, for the simple reason that this is not its purpose.

For scientists, entrepreneurs and, above all, investors, it would not make sense to invest time and money on a synthetic brain that is the same as a human one. You might as well hire a person. With increasingly more data produced around the globe, what does make sense is investing to have something that can extract information from a huge amount of data, and use it. Something humans find it very hard to do.

As Berners-Lee wrote with Hendler and Lassila in 2001, "The Semantic Web will enable machines to comprehend semantic documents and data, not human speech and writings." Computers will be able to take bookings, but they won't understand what a hotel is.

The fact that a software uses our own language does not necessarily mean it also has the same internal representation of reality. We have a human representation of reality, based on our history and evolution, computers don't. The advantage of neural networks is that they learn to do "the right thing" without needing an operator to program the logic, but this is a weakness too. A computer for which the right thing is making paper clips, won't stop until everything is clips (Bostrom 2003). A computer trained to win at Go doesn't have the necessary sensibility to teach the game to a child. A computer trained to generate revenue won't stop if a few people die of infection, or if after thousands of years of war and just one century of peace, a continent, Europe, risks falling into nationalist chaos. Once again.

Earth today, as Gauss imagined, has been almost all completely wired and connected. Soon, probably, there won't be any isolated area. Every thing, and every person, everywhere, will continuously be connected to the Internet thanks to a network of satellites.[23] The amount of information managed by Internet is already beyond the scope of *Homo sapiens*, but it will soon explode.

There is not only the statistical certainty, considering the evolution of life in the past, that the cognitive abilities of this meta-organism will exceed those of every individual, but also the logical certainty: this book was written precisely to explain

---

[23] "Satellites may connect the entire world to the internet", The Economist, December 8, 2018

why intelligent systems must, at a certain point, aggregate into a system able to process more information.

If the human organism has reached its information processing limit, the only thing it can do to survive is to create information networks of human organisms, and start processing the information "outside" the single element. *Homo sapiens* started doing this a few million years ago when some kind of language was introduced, but that has now become more extreme, with the introduction of a global neural network.

The emergence of the human meta-organism gives us *sapiens* the feeling, as Harari (Atlantic 2018) rightly says, that we are part of society but don't have a real role. Like neurons of the brain, we're increasingly more part of a meta-system, which protects and feeds us, but are less free to learn, to process information, and insignificant in relation to everything else. Just as neurons learnt to communicate with only a few signals, some *Homo sapiens* recently abandoned a language of complex syntax to start using smiley hearts and hashtags.

Our ability to adapt lets us transform ourselves into organisms that can communicate quickly, but are less able to process information. To become cells of this meta-organism, working readily for the survival of the same, without knowing why. In this meta-organism, like the cells of a complex organism, safety increases while freedom becomes a thing of the past.

More and more, the medium is the message: the ability of software to associate words with events in our lives, to use our language as a medium, is mistaken for intelligence. This does not mean, however, that conversational User Interfaces, so called chatbots, are not going to be "the next big thing" after web sites, blogs (web 2.0), and social networks. They most probably will be.

The risk is that *sapiens-sapiens* communication will become a thing of the past, and *machine-machine* communication will be the driving force of the meta-organism. But those machines will still have absolutely no real understanding of our reality. Can we really put our lives in the hands of such systems, even more if their only mission is increasing revenues?

Until today, scientists, with all their weakness but thanks to their obsessive quest for rationality, have helped *Homo* become the dominant species on earth. Whoever started mastering fire ("The Greatest Ape-man in the Pleistocene," as Roy Lewis' (1960) masterpiece title was translated in Italian), Aristotles, Galileo, Newton, Enrico Fermi or Tim Berners-Lee, you name it – their first goal has never been ruling the world or make a fortune.

But they produced tools whose power could be easily exploited by second-class scientists: almost all founders of today's big tech have started their careers as scientists. Bertrand Russell (1971) used to say that "no transcendent ability is required in order to make useful discoveries in science". But, he added, the person "of real genius is the person who invents a new method." Inventing a new medium of communication, the World Wide Web, is genius. Exploiting it through the PageRank is brilliant.

Indeed, it is relatively easy to make an atomic bomb too. But because governments recognise the danger of atomic bombs, access to materials that can be used to make those weapons is controlled at a global level, to minimise the risk of a nuclear catastrophe. Perhaps it might soon be wise to control the development of Artificial Intelligence too.

So-called *Artificial* Intelligence might actually help us live a better life. For this though, we need scientists who are more architects than engineers; artists, or sociologists, rather than technicians obsessed with the search for optimisation.

If things continue the way they are going, the race to create a global neural network will probably be very similar to a new form of nuclear energy. A technology which can improve the quality of life by producing cheap electricity with a low environmental impact, but that was initially used to kill millions.

The difference though compared to nuclear energy is that, those who invented Artificial Intelligence have less and less control over it, and the technology is becoming a necessity, much more than nuclear power years ago. It can't be simply swept under the carpet and forgotten. The use of Artificial Intelligence, as the emerging fascism reminds us, risks being an autoimmune disease – the defence systems attacking the organism itself, rather than its enemies – which could lead the human race to unleash the last fatal attack: against itself.

# Appendix 1: More on Networks and Information

## Physicists Don't Like Networking

Physicists love interactions, but only if no people are involved.

Isaac Newton's gravitation, the first great theory concerning the action of matter on matter, revised by Albert Einstein in his general theory of relativity, precisely describes the evolution of macro-aggregations of matter (stars, galaxies). Particle physics on the other hand, in the so-called *Standard Model,* a mathematical model established during the third quarter of the past century, precisely describes the behaviour and structure of subatomic particles, based on weak, strong and electromagnetic interactions.

Despite the fact that physics is still far from coming up with a theory of everything – the universe is full of dark matter, the origin of which we cannot yet explain, and the origin of the universe is still in the realm of metaphysics – many of the mechanisms that govern the fundamental interactions of matter have been modelled in an incredibly accurate way.

Yet we are nowhere near fully understanding the emergence and working of a living organism. We can describe how quarks bond to form an atomic nucleus. We know that atoms contain protons, neutrons and electrons, and atoms bond to form more or less complex molecules – but we don't know how billions of atoms can spontaneously bond into protein chains, that are in turn organised in the form of biological cells and therefore complex organisms, like our own.

There is a branch of physics that deals with the origin and evolution of aggregated systems – the theory of networks – but until the end of the last century not one physicist and only a few mathematicians had studied it. For the entire century the theory of networks was used mainly in "soft science"[1].

---

[1] The separation science in "soft" and "hard" science reminds of the physicist Sheldon's disdainful definition of biology in 'The Big Bang Theory': "…biology: that's all about yucky, squishy things".

© The Author(s) 2020
M. Alemi, *The Amazing Journey of Reason*, SpringerBriefs in Computer Science, https://doi.org/10.1007/978-3-030-25962-4

Sociologists and psychologists started studying social networks, on the basis of "it's a small world!". "A cliché to be uttered at the appropriate moment of recognizing mutual acquaintances", wrote the psychologist Stanley Milgram (1967) in his article entitled "The Small World Problem". Milgram, a social psychologist, published the results of an experiment that proved the level of separation between two Americans was around 5 – in other words two randomly chosen Americans looking through their friends, on average, found they were the friend of a friend of a friend … and so on to the fifth level.

In his article on the subject Milgram apologised for not having found a mathematical model that explained how social networks could be so closely connected. Someone else took a shot at it. In 1997, 116 scientific articles quoting Milgram's work[2] were published. They were all in sociology, psychology or anthropology publications.

## Abby Normal Distributions[3]

Another phenomenon, apparently not connected to the theory of networks, seemed to be of little interest to physicists and other "hard scientists": many observables in nature are not distributed in a Gaussian or "normal" way. Many physicists and mathematicians assume that when something is measured, most of the results oscillate around an average value, and we observe exponentially fewer results the further we get from the average. People's height for example, follows Gaussian distribution: 68% of Italian women are 155 to 170 cm tall, but only three in every thousand are over 177 cm[4]. No one is 10x taller than average or 1/10th average human height.

But observables that follow a Gaussian distribution, like the height of humans, are not … the norm[5]. Many observables follow the "Pareto-Zipf-Levy-Mandelbrot" distribution – called by whichever of the names of these scholars you prefer, none of which was a physicist. It's also simply referred to as "power law".

Because it was initially studied by Italian economist Vilfredo Pareto, we will refer to it also as the "Pareto distribution". Pareto (1896) realised that the distribution of wealth in Europe didn't follow the Gaussian distribution. He did that tracing

---

[2] Search made on scholar.google.com the 7 June 2018.

[3] Dr. Frederick Frankenstein: Whose brain did you put in him?

Igor: Err… Abby something…

Dr. Frederick Frankenstein: Abby who?

Igor: Abby… Normal. Yes that's it, Abby Normal!

Dr. Frederick Frankenstein: Are you saying that you put an abnormal brain in a 7 foot tall, 54 inch wide gorilla!? (Brooks, M. (2000). Frankenstein junior. Twentieth Century Fox Home Entertainment).

[4] https://tall.life/height-percentile-calculator-age-country/, checked the 11 June 2018

[5] Gauss came up with normal distribution in his theory of errors. It says that the measurements of the same sample follow Gaussian distribution because it is impossible to take perfect measurements. It is a property of the observer, not of the system.

a graph showing the number of people "N" whose wealth is "x" or more, for each value of "x"[6]. He noticed then that for all the countries he had data on, the curve of the graph could be represented as a relation

$$N \sim 1 / x^a$$

With "a" measured as between 1.35 and 1.73. Interestingly, Pareto (1909) also attempted to explain the *skewness* (asymmetry) of the wealth curve, in other words the long tail on the right (there are some people who earn 10 or even 100 times more than the average salary). His hypothesis was that there are no limits to the income of an extraordinary person, while a person with negative qualities cannot earn less than the subsistence level. Unfortunately, this is a purely qualitative reasoning which might explain the asymmetry, but not why the Pareto distribution emerges[7].

In finance, the Pareto curve hasn't been used as much as one might expect. For example, the Basel II Accord of 2004 accepted a Value at Risk based on log-normal distribution (Basel 2004) therefore unable to predict, in the 2008 financial crisis, what seemed to be extremely rare events, nicknamed by Nassim Taleb (2007) Black Swans.

In the academic field some scholars started asking themselves why the Pareto curve occurred. In 1963, the mathematician Benoit Mandelbrot (1963), who would become the father of fractal geometry, noted there was nothing normal in the variations of the price of cotton: there is relatively little difference in the majority of cotton prices, as in the salaries in Pareto's studies. But every now and then there are huge variations, just as every now and then someone gets extremely rich[8] (for a review see Fama 1965).

The Pareto distribution also appears in the "Zipf's law" after the linguist George Zipf. Zipf (1932), analysing word frequency in various corpora in English, Latin and Chinese, observed that the distribution of the frequencies followed the trend $1/x^a$ with a=2, and also that by putting the words in order from the most frequent to the least frequent, the product *rank•frequency* remained constant, in other words

$$Rank \sim K / x^{a-1}$$

---

[6] By "or more" Pareto shows that he is measuring the cumulative distribution, in other words the sum of all those whose income is greater than x.

[7] In fact – Pareto's reasoning led to the introduction of the "log-normal" curve, in which the factor applied to the mode is normally distributed (half of the mode has the same probability as double, a third as triple, and a fraction as the infinite). In finance in fact, fluctuations in the prices of shares were initially modelled using the normal curve on the basis of the pioneering works of Louis Bachelier (1900), and the log-normal was adopted in the 1960s

[8] Mandelbrot, basing his studies on the work of his master Paul Lévy, went further, proving that "the tails of all non-Gaussian stable laws follow an asymptotic form of the law of Pareto"

Where $K$ is a constant and $a$ is 2. The second representation is equivalent to the one used by Pareto –the cumulative distribution, but with the abscissas in place of the ordinates, therefore using the inverse of the Pareto index[9].

## Networks and Power

Pareto distribution and social networks are strongly related. In social networks, the importance of the individuals, measured by number of contacts, follows Pareto.

Most people have a few dozen connections, but there are some individuals who are hyper-connected, with thousands of connections. Choosing one person at random, and following a path through their contacts, it's highly probable that we'll end up going through one of these social hubs.Their existence makes the network so hyper-connected. If we want to send a message to a stranger (as in Milgram's original experiment) obviously we'd contact someone who, as far as we know, has a lot of contacts. And this person in turn, with many contacts, will have someone in their contacts with even more contacts, so the message will be delivered in just a few steps thanks to these "social hubs".

One possible mechanism responsible for the development of these networks was first proposed by Reka Albert and Albert-Laszlo Barabasi (1999): incremental attachment. In their articles, the two physicists proved that the assumption that each new individual who joins a network will connect to another individual with a probability proportional to the number of their connections, is sufficient to spontaneously create a small-world network.

Inequality in networks (the fact that a few nodes are super-rich and most nodes are poor) is linked to how connectivity evolves: those with most connections gain even more connections, and in doing so the network grows to be highly connected. Our society is a network of economic relationships, and is no wonder that the rich gets richer.

## The Emergence of Small-world Networks

Just before (Albert and Barabasi 1999) was published, two mathematicians, Duncan Watts and Steven Strogatz (1998), proved that to create a small-world network all you have to do is take an ordered network, like a lattice, and introduce a pinch of chance, severing a few links and recreating them in a random way. The work by

---

[9] It can also be proven through integration that Pareto's cumulative distribution index is in fact that of the same distribution less one, therefore the Zipf distribution index must be $1/(a-1)=1$, as observed.

Watts and Strogatz is fundamental as it proves that not only social networks, but also nervous or electrical systems are small-world, and these networks are all highly connected and have a high clustering coefficient, in other words "my friends are friends of my friends".

In practice, after a century, the theory of networks has managed to prove that systems made up of elements that interact also over great distances, such as nervous systems, the web, social networks, etc. have the following characteristics:

1. The number of connections follows a power law, in other words the probability that the $n^{th}$ most connected element receives a new connection is proportional to $1/n^a$, where $a$ is bigger than 1.
2. Point 1 may be the result of a growth mechanism that makes networks:

   a. highly connected
   b. highly clustered

Almost all the articles quoted above are considered cornerstones of the theory of networks. With so many physicists on the subject, one might think that the concept of entropy is essential in this theory. It isn't: even in (Albert and Barabasi 2002), called significantly "Statistical Mechanics of Networks", the word "entropy" is never used.

## Entropy of Networks

First of all, let's see how we can define microscopic and macroscopic description in a network.

The microscopic description is as usual the one where each node (as the elements of networks are known) can be identified for its own properties, like color or a label "A".

The macroscopic description derives from the role each node plays inside the network, or, more formally, how the topology of the network can identify (univocally or not) the node.

Let's consider a very simple network: a triangle.

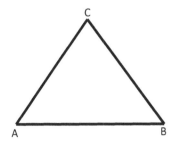

The microscopic description of the system is one that identifies every node with the letters A, B, C. The macroscopic description instead describes each node as "the one with two connections" –not useful indeed.

But if, instead of a triangle, we consider a network with three different nodes, like this:

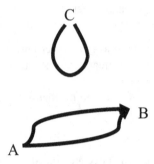

the description we can use for each node thanks to the properties of the network ("the node with two incoming connections", or "the node with the loop", or "the node with two outgoing connections") has the same uncertainty as the microscopic description.

Thanks to our skills in analysing networks, we can say that the difference between the microscopic and macroscopic description (Boltzmann's entropy as redefined by von Neumann) in the first case is:

$$\text{entropy}\left(\text{triangle}\right) = \text{uncertainty}\left(\text{macroscopic}\right) - \text{uncertainty}\left(\text{microscopic}\right)$$
$$= -\log_2\left(\frac{1}{3}\right) - 0 = 1.58\,\text{bit}$$

While in the second case is 0 bit.

For those interested in some mathematical formalism, in the triangle all the permutations of the nodes (6 possible) produce the same network. If we switch "A" and "B", remaining in a topological context so without considering the position of the nodes, we obtain the same network.

This allows us to define entropy in a network with N nodes as:

$$S = \sum_{i=1..k} \frac{n_i}{N} \log_2 \frac{n_i}{N}$$

where each $n_i$ is one of the $k$ groups of nodes that can be permuted without changing the macrostate.

In this network, taken from Mowshowitz (2009), a publication that provides a mathematically more formal description of entropy in networks:

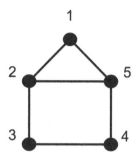

we can switch nodes 2-5 and 3-4. So there are three groups of interchangeable nodes: {1}, {2, 5}, {3,4) and entropy is

$$S = -\frac{1}{5}\log_2\frac{1}{5} - \frac{2}{5}\log_2\frac{2}{5} - \frac{2}{5}\log_2\frac{2}{5} - \frac{2}{5}\log_2\frac{2}{5}$$

The mathematical formalism in (Mowshowitz 2009) can be used to calculate the entropy of each network on the basis of the so-called adjacency matrix, a matrix in which each element i, j indicates the existence or not of a link between nodes i and j.

## Barabasi-Albert small-world networks

It's not always possible to find a group of interchangeable nodes in a network using a computational operation. For this reason it's better to simplify the entropy formula, and instead of considering all the network topology, merely limit the operation to a few properties of the same topology, such as the number of connections of the nodes. The groups of interchangeable nodes in the topology are replaced by the groups of nodes with the same number of connections: all the nodes with the same number of connections are indistinguishable.

In this case the entropy of the network is calculated using degree distribution. In a Barabasi-Albert network, in which the probability that a node receives a connection from a new node is proportional to the number of connections $k$ it already has (incremental attachment), the degree distribution comes close to the curve $1/k^3$ (Albert and Barabasi 2002, sec. VII B).

As $\sum_{k=1}^{\infty} k^{-3} \sim 1.2$ [10], a network of $N$ elements will have approximately $N/(1.2 \cdot k^3)$ nodes with $k$ connections.

---

[10] The series is the so-called "Euler-Riemann Zeta Function", one of the most famous (and most studied) mathematical functions. The Zeta function for z=3 converges with "Apéry's constant", an irrational number around 1.2

In a Barabasi-Albert network of 10,000 nodes, the probability of having a node connected to all other nodes, i.e. a node with 10,000 connection, is practically zero. But in the brain, which has 100 billions nodes, the probability for a neuron to have 10,000 synapses is far from zero (Gazzaniga 2009, page 366).

Incremental attachment gives us a simplified, but faithful, model of reality: a network that grows with this mechanism has a low degree of separation and observables distributed according to power-law.

Barabasi-Albert networks help us understand many network phenomena. Let's take a look at intelligence. So-called intelligence quotient (IQ) tests for example, are considered standardised by definition, with an average IQ of 100 and a standard deviation of 15. Those who score 100 are considered normal, over 115 (a standard deviation) particularly intelligent, less than 85 not very. But if we consider our own personal experience, it's hard to accept that intelligence follows such a bell curve trend. Just as some rich people may feel poor when compared to others who are richer (imagine your richest friend how would they feel when compared to Jeff Bezos – even if they are the hundredth on richest people ranking), in the same way even the brightest people in a degree course might feel like nonentities when compared to the greatest minds of all time – those four or five who changed our understanding of the world.

The reason for this, as confirmed by recent studies (Aguinis 2012), is that the intelligence tail, as vaguely as intelligence can be defined, is extremely long and not at all Gaussian.

The importance of the Barabasi-Albert model is that it simulates the fact that, the larger the network, the more exceptional elements –so called outliers– there are. Many other parameters plays an important role, but a bigger society produces outliers more easily than a small one. The "many other parameters" might well be how well connected the society is, how valuable the information it stores, and more. But if we apply the same Albert-Barabasi growth process to two societies, the smaller one will have less outliers.

Studies on different IQs in various communities, such as in (Herrnstein 1995), eloquently entitled "The Bell Curve, Intelligence and Class Structure in American Life", can draw some strange conclusions if they fail to consider this simple growth process. A simulation shows that an observable generated randomly on the basis of a power-law with an index of 2 and a minimum value of 1 in two communities, one of 200 million inhabitants and another of 30 million inhabitants (approximately the "Caucasian" and African-American ethnic groups analysed in Hernstein's book) generates outliers in the first community with an "intelligence" value 30 times higher than the second. These are randomly generated numbers, and take into account just the size of the two communities, but as the first community is larger, and the tail is a lot longer, it will have outliers that cannot be found in the second[11].

In small-world networks, "size matters".

---

[11] It is an extremely simplistic model: the Jewish community in Europe has always been a sub-community, but it has produced quite a few notable geniuses. The importance of teaching in the jewish community, for instance, might have prevailed over the size. Geniuses like Paul Erdős considered the time they dedicated personally to teaching gifted children to be just as important as time spent on research (Hoffman, 1998). Thanks to Tito Bellunato for pointing this out.

## Why Power Law Appears in Networks

In "Complexity in Numbers" we discussed Kolmogorov's complexity, in other words how easy it is to come up with a formula that faithfully describes network topology.

Shannon's formula

$$H = \sum_i - p_i \cdot \log_2 \left( p_i \right)$$

was in fact proposed as part of communication theory. It was not dealing with complexity, nor with meaningfulness of a message.

Shannon's communication theory considers the question: What do I need in order to encode a message (whatever the meaning), send it, and be sure the person who receives the code can decode said message?

In practice, the fact that the letters, and the bigrams (two letters), trigrams, etc. don't occur with the same frequency, helps us decode the message. For example, if you receive "WHO" and then "FEVER", you might imagine that the message had been disturbed by interference and was originally "WHO EVER" (Shannon 1948). The entropy of a language in which the probability of a character is simply the reciprocal of the number of available characters is greater than for a language in which the probability of each character is different, and what's more depends on which characters precede that particular character. The same thing is true for words.

In English, for example, entropy was originally calculated by Shannon using consecutive approximations, in other words considering the probability of observing each letter as being the same ($F_0 = \log_2 26 = 4.7$ bit per letter), then considering the actual frequency $p_i$ of each letter ($F_1 = -p_a \log_2 p_a - p_b \log_2 p_b \ldots -p_z \log_2 p_z = 4.14$ bit) then considering the probability of one letter on the basis of the previous one, ($F_2 = 3.56$ bit) and so on (Shannon 1951).

Next Shannon approximated the frequency with the Zipf's distribution, calculating an entropy of 11.82 bit per word (Shannon 1951), soon after corrected to 9.74 by Grignetti (1964).

What do all these "bit" mean for the English language? For example, if we were to consider the statistics of bigrams, we wouldn't need 7 bits to encode each letter, as we do with ASCII code. $F_2$ says that 4 bits are more than enough. We use 7 bits so that the message is redundant, and we can understand it also in the presence of noise. Smartphones' "predictive keyboards" (such as SwiftKey) are a perfect example. To input a word, one or two letters are enough, because the smart keyboard also takes into consideration the previous words, and can guess that after "WHO" there will be "EVER".

You'll notice this when speaking too; it's that redundancy that helps us understand when there's a lot of background noise. Without redundancy, each of the 100,000 English words (the total number of words used in Grignetti 1964) could carry

$$\log_2\left(N_{words}\right) = \log_2\left(100{,}000\right) = 16.6\,bit$$

of information. Because of redundancy this is reduced to 10 bits.

It is a bit surprising that the frequency of words follows the Pareto distribution. As noticed by Easley (2010), "when one sees a power law in data, the possible reasons why it's there can often be more important than the simple fact that it's there".

The reason of such a distribution is that Pareto gives us "the best word-by-word coding". Mandelbrot (1953) proves that a function that generalises power-laws minimises the cost (i.e. the amount of bits) required to send the message with a certain level of redundancy. Soon after, Simon (1955) reached the same conclusion on the basis of various assumptions (and acknowledging the ingenuity of Mandelbrot's approach).

In the Barabasi-Albert mechanism this does away with the "metaphysical component" of preferential attachment: while we can accept that it may be true for social networks and the web, can a neuron really "know" it should preferentially connect with neurons that have many connections? And what about the protein in a cell? Is there such a thing as a prime, absolute principle on which life is based, that drives these components to aggregate in such a way as to create complex organisms?

It would appear to be more likely that the complexity of networks, discernible from the fact that they are organised and grow as if following a predefined design, is due to the fact that networks, like language, are just a support on which information is stored and the emergence of power-law is the effect of storing information, done in a redundant way, but with minimal effort.

## Code snippet: entropy of scale-free networks

```
import math
import numpy as np
import collections
def network_entropy(a, node_thousands, stat=10):
"'Entropy of a network of 'node_thousands'*1,000 ele-
ments based on an observable following a Pareto distri-
bution with exponent a+1 (see https://docs.scipy.org/
doc/numpy-1.14.0/reference/generated/numpy.random.
pareto.html)
"'

def par(a, node_thousands):
ca = {} # counter - how many nodes have a certain distri-
bution value
for t in range(node_thousands):
try:
steps = int(node_thousands/100)
```

```
if t % steps == 0 and t != 0:
pc = t // steps
print("INFO: %s per 100 done (%s)" % (pc, t))
except ZeroDivisionError:
pass
w = (np.random.pareto(a, 1000) + 1) # get a random node
with its distribution value
for i in w:
ca[round(i, 0)] = ca.get(round(i, 0), 0) + 1
ent = 0
for v in ca.values():
ent += v/float(node_thousands*1000) * math.log(v,2)
return ent, ca
e = []
for i in range(stat):
ent, ca = par(a, node_thousands)
e.append(ent)
return np.mean(e), np.std(e), ca
```

# Appendix 2: Math and Real Life

The book used the concepts introduced chapter 1 (information, energy, complexity) as keys to follow the emergence of life and its evolution. There are two questions which one might ask though:

3. Did scientists come up with things like the mathematical theory of information, entropy, energy, just for fun, and then engineers found practical applications?
4. There was a lot of writing about complexity. Simons' definition (the whole is bigger than its parts) makes clear when a system can be called complex. But why do we say that certain structures are complex, and others are simple?

Here we have then a brief history on how information theory was used by its creators to start making some money, plus an explanation of why equally probable series of digits are seen by us as "simple" or "complex".

## Two Scientists Playing Roulette

It's fair to believe that no living being minimises the uncertainty it has on a system using mathematical formulas, and so becomes able to extract energy from such system. There are exceptions though: scientists and hedge funds. Edward Thorp, physicist, gambler and creator of the first hedge fund, is the person who did that, together with Claude Shannon, the father of information theory and creator of what we have called the "uncertainty formula".

In "The Invention of the First Wearable Computer", Edward Thorp (1998) tells how he and Claude Shannon built a small device that predicted which of the roulette wheel's octants the ball would fall into with sufficient precision to be sure, in the long term, that they would leave the casino with more money than when they came in. "Sufficient precision" meant for Thorp and Shannon knowing that the probability of the ball stopping into one particular octant on the roulette wheel is higher than 1 in 8.

© The Author(s) 2020
M. Alemi, *The Amazing Journey of Reason*, SpringerBriefs in Computer Science, https://doi.org/10.1007/978-3-030-25962-4

If Thorp and Shannon could have actually predicted, with the laws of mechanics, in which octant the ball would fall with absolute precision, they would have reduced their uncertainty from $\log_2(37) = 5.2$ bit to $\log_2(8) = 3$ bit, improving their winnings from an average loss of

Loss(average roulette player) = $36/37 - 1 = -2.7\%$

To an average win of

Win(Thorp & Shannon) = $36/8 - 1 = 350\%$

Of course, the difference between losing and winning is more than just a question of decreasing uncertainty – there are other variables to consider too. In a casino where the payout for a win is multiplied by a factor of 7 instead of 36, the Thorp-Shannon system reduces losses, but doesn't give a win. In a similar way, if the device required greater investments in terms of components, electrical energy and work than the payout, the balance at the end of the day would be negative anyway.

(In fact, the device wasn't a success due to research and development costs and continuous faults, but the mathematical technique – the information gain – was put to good use by Thorp to set up the first hedge fund in the history of finance).

We could study the analogy between a casino and thermodynamic systems in greater depth by defining a Boltzmann constant and temperature for a roulette, but this isn't inside the scope of the book. The fundamental concept is:

1. You have to invest energy to be able to extract energy. In the above experiment, Thorp and Shannon worked for almost a year on their device.
2. There is a direct relationship between the information – or reduction of uncertainty – we have on a system and the amount of energy we can extract from that system. Having information on a system (in other words not being in a situation of the greatest uncertainty) is a necessary condition, but alone it's not enough to extract energy.

Note that the information we have defined is not found in Shannon (1948). As explained by Weaver: "The word information, in this theory, is used in a special sense that must not be confused with its ordinary usage. In particular, information must not be confused with meaning" (Weaver 1949). Shannon's information is statistical, syntactic information, but the information isn't considered in terms of reducing the uncertainty of the message (e.g. concerning a possible winning number or weather forecasts, as in Eco 1962).


## Complexity in Numbers

John von Neumann's definition of entropy can be used to explain why we consider a sequence of numbers "ordered" or not.

If a message was received from outer space that was interpreted as 3.1415... mathematicians would have no doubt whatsoever: it came from an intelligent, highly evolved life form that sent us $\pi$ ("pi", the ratio of circumference divided by the diameter).

For mathematicians, a huge amount of information can be communicated by $\pi$, and from Archimedes to the present day this information continues to grow: it's almost as if every formula drawn up since the nineteenth century is based on $\pi$.

As humanity needs the finest mathematical minds to define both the expansion of the number and its use, it's also fair to say that $\pi$ is highly complex, but its complexity has been tamed through formalism, identifying this quantity in a fast and precise way.

Still, the series of digits in $\pi$ on its own is not special. The sequence of digits in $\pi$ is no more or less improbable than any other sequence, including 0.111111111... or even just "1". Generating random digits from 0 to 9, it's improbable the result will always be 1, or 0, but it's just as improbable that the result will be the decimal digits in $\pi$ in perfect order. The same goes for the square root of 2, which was "discovered" long before $\pi$.

Furthermore, we know that periodic sequences – rational numbers – represent just an infinitesimal fraction of non-periodic numbers like $\pi$ or Euler's number $e^1$ – so-called transcendental numbers.

So, there are a few natural numbers (1, 2, 3...), just as many rational numbers (fractions) and algebraic numbers (numbers that can be expressed as the root of a polynomial equation with integer coefficients) and infinitely more transcendental numbers. Complexity increases in a similar way: natural numbers are not very complex, transcendental numbers are much more complex.

Complexity means how well we know these numbers. We know natural and rational numbers from head to toes. Algebraic numbers are harder to represent, but – by definition – they can be reduced to simple formulas. But representing transcendental numbers is really hard, that's why we are familiar to only a few of them although they are infinitely more.

Complexity has been "tamed" for some transcendental numbers in the sense that as mathematics progressed, the entropy as defined by von Neumann we associate with $\pi$ has constantly diminished: the difference in uncertainty between the macrostate (our description of $\pi$ using a symbol) and the microstate (the actual number $\pi$) is "virtually minimal". We must of course say "virtually" because in any case we only know a few trillion of the digits in $\pi^2$, which we know is only a tiny part of the actual number.

The above is in line with Kolmogorov's definition of complexity: a series is complex if it is difficult to define with a formula (it took us thousands of years since we started studying mathematics to accept the fact that irrational numbers like the square root of 2 existed, and even longer for transcendental numbers), non-complex in the opposite case (such as a natural or rational numbers). Therefore transcendental numbers are by far the most complex, followed by non-transcendental, irrational and rational numbers.

---

[1] See http://en.wikipedia.org/wiki/E_(mathematical_constant)

[2] See "y-cruncher - A Multi-Threaded Pi-Program", url: http://www.numberworld.org/y-cruncher/

This leads us to conclude that measuring Shannon's entropy in a series of digits considering only the frequency of said digits is not always a good instrument for acquiring information content. If we do that both square root of 2 and $\pi$, along with infinite other sequences, meet all the requirements to be considered a casual series of numbers (in other words the statistical non-predictability of a number on the basis of the ones before it). They would seem an informationless sequence of digits.

Still, square root of 2, and even more $\pi$ and e, seem to carry a good amount of information about nature!

Using Shannon's formula to measure the information content is misleading, or just plain wrong, for two reasons. First, because Shannon introduces entropy for communication, and says nothing regarding the information content. Second, because if we measure complexity using Shannon's entropy and statistical correlations, we might be unable to tame complexity, and appreciate the inherent structure of a system.

# Appendix 3: How Artificial Neural Networks Work

At the end of the last century it would have been technologically unthinkable to keep files on billions of individuals and observe their behaviour in order to produce tailor-made content on the fly. Even in 2010, the leading technology used in big data analysis was powerful but still relatively slow. Hadoop (White 2012), the open source software created following the specifications of Google MapReduce (Dean 2004) and GFS (Ghemawat 2003) was the standard for the (few) companies processing big data, but it could only be used for offline data analysis.

Today, on the contrary, not only the data but also the almost instantaneous analysis such data are becoming commodities. So what technology made this change of paradigm possible?

Italian Renaissance polymath Gerolamo Cardano, with his passion for gambling, was the first person to introduce the concept of probability. Probability calculation can be considered a search for a logical instrument used to reduce uncertainty: from Cardano to Thorp, who four hundred years later used probability in finance, it gives those who use it an advantage based on intelligent data analysis.

But as we saw, it's not just gamblers and hedge fund investors who want to reduce uncertainty: the principal task of every intelligent system, living ones included, is to predict the evolution of other systems, from which energy can be extracted thanks to an information advantage, with the lowest possible uncertainty.

There was a minor revolution in probability at the dawn of the new millennium. In the last years of the 20th century, the frequentist approach was still the dominant paradigm. In order to calculate the probability of a coin toss coming up heads or tails, I toss a coin a high number of times and note the result: as the coin in asymptotic terms appears to land heads and tails with the same frequency, I can deduce that the probability is 50%.

In recent years however there has been more widespread use of the Bayesian approach, mentioned briefly in chapter 2 ("The importance of scientific revolutions"). Probability is no longer calculated in an agnostic way: I don't need to toss the coin to imagine that a flat cylinder made of metal, subject to various forces, can reach a minimum potential energy state with either one side or the other facing up,

M. Alemi, *The Amazing Journey of Reason*, SpringerBriefs in Computer Science, https://doi.org/10.1007/978-3-030-25962-4

which is a complicated way of saying that intuitively the probability for heads or tails are the same.

In short, I already have an idea beforehand of what might happen. An idea based on experience and past models and that I can, if necessary, adapt to new observations.

In practice, in the Bayesian approach, the probability shows only "how certain I am that something will happen". The Bayesian definition is looser than the frequentist one. This is a good thing, because the frequentist definition makes it difficult to calculate probability when experiments can't be repeated as often as you'd like. One extremely practical use is in weather forecasts, where the only definition that can be used is the Bayesian one: I can't observe tomorrow's weather 100 times to express the probability that a weather forecast is right or not.

Weather forecasts are a fine example of a problem solved by using the Bayesian approach and artificial neural networks, the technology that made a new revolution in data analysis possible. To forecast the weather, I might have thousands of input parameters (e.g. pressure, humidity and temperature from various parts of the world, with readings taken every hour).

In exactly the same way as the ancient Sumerians, who wished to predict the motion of planets without using a mathematical model that explained said motion (we had to wait for Newton to do that), I could hope to forecast the weather by simply creating an algorithm, a series of formulas, that learns from the past, but isn't based on a logic model.

It might seem easy, but it isn't.

I have thousands of input parameters, and all of them are used to forecast the weather in a non-linear way. Non-linear means the temperature tomorrow may initially increase in a proportional way to the temperature today (if today is hot, tomorrow will be hot), but suddenly drop if it rises a lot (if today there's a hot spell, clouds will form, and tomorrow will be colder).

When one variable is related to another in a non-linear way, chaotic behaviour may emerge. In mathematics, chaos means highly unpredictable behaviour, and this is what these relationships produce. Let's say that, after normalising the temperature range between 0 and 1, one discovered that tomorrow's and today's temperature are linked by an apparently simple relation

$$T_{tomorrow} = r \cdot T_{today} \left(1 - T_{today}\right)$$

For r > 3.57 it would be practically impossible to forecast the weather in one week's time on the basis of the weather today. The above formula, the logistic map[1], is in fact one of the best known examples of deterministic chaos, due to the non-linear dependency between the value of a variable (temperature in our example) now and at the previous time of observation.

---

[1] All books on deterministic chaos include the logistic map. There's an excellent Wikipedia article: https://en.wikipedia.org/wiki/Logistic_map

It seems strange, considering the weather of one particular day is determined univocally on the basis of the weather of the previous day, but these systems are highly unstable and even a tiny change in the input data produces an evolution in the system in a totally different direction.

As the precision of every measuring instrument, including thermometers, is finite, input data never represents extreme conditions with infinite precision. This small difference between real conditions and theoretical conditions increases in time.

Not surprisingly, the chaos theory was discovered in meteorology along with the dependence of the initial conditions called the "butterfly effect": the flap of a butterfly's wings in Brazil can set off a cascade of atmospheric events that changes the weather all over the planet in a way which wasn't originally expected (Lorenz 1963, 1972).

Predicting the evolution of a highly non-linear system dependant on thousands of parameters, like the planet's weather, seems impossible. Even if we had a complete model, we could never measure all the input variables with infinite precision, for example by observing all the butterflies in the world.

As a model based on the laws of physics does not in any case help a great deal to forecast medium-long range weather, I might think I could create a generic model, that isn't based on the laws of physics. A highly non-linear model, in the hope that I could adapt it to the real situation. This is what our brain does when it instinctively predicts something: it tries to predict without creating mathematical models, just with "circuits" in the brain able, with certain input, to provide an output that can be interpreted as a prediction.

It's something mammals find it easy to do. The behaviour of another person is no less complex than the evolution of the weather. Nonetheless, we interact easily with others thanks to our ability to predict their behaviour. To do so we do not build a mathematical model of our acquaintance's brains.

In a similar way, to predict the evolution of the weather system I don't try to understand how it functions with a logical-deductive model, but connect a high number of variables in a non-linear way, in the hope that this network can recreate, approximately, the behaviour of the system I wish to modelize. I create an internal structure that reproduces the behaviour of the external one, but with different mechanisms.

This approach would appear to be of an indefensible complexity.

Let's imagine a simpler task to the one meteorologist are faced with: we wish to predict the atmospheric temperature tomorrow on the basis of the temperatures taken by a network of a thousand thermometers all over the world.

What we do to create the non-linear model is choose a function, obviously not a straight line, using the thousands of temperatures taken as the input. For example, a function that uses the weighted sum of all the variables as the input, the result of which is 0 until the weighted sum is less than 7, and increases to 45 degrees for values of over 7:

$$f(p_i, \alpha_i) = \begin{cases} 0 & \text{if } \sum \alpha_i \cdot p_i < 7 \\ \sum \alpha_i \cdot p_i - 7 & \text{if } \sum \alpha_i \cdot p_i \geq 7 \end{cases}$$

The output of this function can be seen as a mathematical neuron, the input of which is the output of all the sensory neurons –the temperatures taken by the thermometers. The function has 1001 parameters: those of the weighted sum ($\alpha_i$), and the threshold (in this case 7), after which the neuron emits a proportional input signal.

This would already appear to be complicated, but we aren't finished with this process yet. We construct another 500 of these functions/neurons, each of which has 1,001 parameters, the input of which is the output from the 1,000 neurons of the first layer. Then we do the same thing with the output of these functions, and construct 100 other functions, connecting the first 500 neurons to 100 new neurons, this time each has 501 parameters for the input functions. Finally, we connect these 100 functions to 50 functions, our output neurons.

This makes in the end 555,650 parameters.

At this point we associate a probability with each of the 50 output neurons. The first neuron expresses the probability that the temperature will be 10 degrees centigrade below zero or lower, the second above -10 and below -9, and so on up to the last neuron which represents 40 degrees above zero or higher.

The solution lies in managing to train this network of functions/neurons so it can predict the output on the basis of the input. In other words, find the set of values for the more than half million parameters that, when a certain input is given, allows the network to predict the actual observation. If we have one year of readings, for each year we'll use the output of the weather stations as input, and we'll choose a set of parameters so the output neuron close to the observed temperature has a high value, and the others will be close to 0.

This operation, in reality, would appear to be not only a crazy idea, but also a useless one too. As Enrico Fermi said to Freeman Dyson: "My friend Johnny von Neumann used to say, with four parameters I can fit an elephant, and with five I can make him wiggle his trunk." There will certainly be thousands of sets of parameters that, given the temperatures of one day, can be used to obtain the temperature taken the next day. We might as well become astrologists and use the position of the planets in the various constellations as parameters: for every past temperature we will find a series of weights to associate with each position of the planets, but without the power to predict the future!

What one does when building artificial neural network is more sophisticated than finding "a set of parameters that can be adapted." One looks for a set of parameters that lets each layer create a representation of the global situation, which can then be analysed using the next layer.

Think of how a neural network (artificial or in the brain) could classify the image of a dog: first it would classify it as the image of a living being, then as an animal, and then as a dog.

To do this, when training the network (in other words adapting the parameters) one enters the photograph of a dog in the initial layer and tell the network that in the final layer the highest value must be that of the neuron associated with a dog.

The neurons are not adapted by changing the values of the parameters one by one in an iterative way, which is a bit slow, not optimal, but feasible. One starts from the last ones, then those in the second last layer, which are the values in the second last layer that let the last represent the animal?, then the third last, and so on.

Because the network is trained starting from the last layer and going back, this algorithm is called backpropagation (Werbos 1974) (Parker 1985) (LeCun 1985) (Rumelhart, Hinton, Williams 1988). The values of the parameters on all the other layers are obtained using the infinitesimal calculus, developed by Newton and Leibniz. The interesting thing is that a dog and a cat are represented in the same way in the first and second layers: the network learns autonomously to classify them in the same "living" and "animals" macro-categories.

We don't know how a neural network could represent global weather, but it is similar, because first the macro-categories are identified (e.g. emergence of an area of high pressure) and then, greater detail is gradually added.

The network, also when forecasting weather, learns to recognise different representations of input data. These go from more general (the first layer, with 1,000 neurons) to more specialised (the last, with the temperatures). The network of functions simplifies the representation as the signal progresses in the network.

Although the topology of the neural network described above is simpler than those used today, the functional principle is the same. Above all, the power of these mathematical instruments is being able to create, as mentioned above, hierarchical representations of reality.

All things considered, there are two things about neural networks that at first glance appear surprising. First, their ability to represent various aspects of reality – they can be used to recognise images when processing natural language. Second, the fact that the more they learn the better they get at learning. A neural network that's trained to categorise images initially needs a lot of data, but as it's used every new entity will be categorised faster and faster.

In exactly the same way, a young child needs time to be able to distinguish a dog from a cat, but an adult immediately knows a lynx is a different animal that is just similar to a cat. A trained network learns faster, because it has already learnt the fundamental thing: how to create representations.

A third thing to consider is that, even if backpropagation (very probably) isn't used by the brain, the neural networks are based on how neuroscientists think the brain works, and in many ways they work like a biological brain.

# References

Achenwall, G. (1748). Vorbereitung zur Staatswissenschaft. Europäischen Reiche. Göttingen, 1748.

Achim, K., Arendt, D. (2014). Structural evolution of cell types by step-wise assembly of cellular modules. Current Opinion in Genetics and Development, 27, 102–108.

Adami, C. (2012). The use of information theory in evolutionary biology. Annals of the New York Academy of Sciences, 1256, 49–65.

Adena, M., Enikolopov, R., Petrova, M., Santarosa, V., Zhuravskaya, E. (2015). Radio and the Rise of the Nazis in Prewar Germany. The Quarterly Journal of Economics, 130(4), 1885–1939.

Aesop. (1994) Aesop's fables. Wordsworth Editions

Aguinis, H., 2012. The best and the rest: Revisiting the norm of normality of individual performance. Personnel Psychology, 65(1), pp. 79–119. (https://cphr.ca/wp-content/uploads/2017/01/True-Picture-of-Job-Performance-Research-Results.pdf)

Aiello, L., Dean, C. (1990). An introduction to human evolutionary anatomy. Academic Press.

Aiello, L., Dunbar, R. I. M. (1993). Neocortex Size, Group Size, and the Evolution of Language. Current Anthropology, 34(2), 184–193.

Albert, R., Barabási, A. L. (2002). Statistical mechanics of complex networks. Reviews of modern physics, 74(1), 47.

Allman, J. (2000). Evolving brains.

Anderson, J. A., Hinton, G. E. (1981). Models of information processing in the brain. In Parallel models of associative memory (pp. 33–74). Psychology Press.

Bachelier, L. (1900) Théorie de la spéculation in Annales Scientifique de l'École Normal Supérieure, 3e Série, tome 17

Bada, J. L., Lazcano, A. (2003). Prebiotic soup—revisiting the Miller experiment. Science, 300(5620), 745–746.

Bar-On, Y. M., Phillips, R., Milo, R. (2018). The biomass distribution on Earth. Proceedings of the National Academy of Sciences, 115(25), 6506–6511.

Barabási, A. L., Albert, R. (1999). Emergence of scaling in random networks. science, 286(5439), 509–512.

Basel Committee on Banking Supervision, 2004. International Convergence of Capital Measurement and Capital Standards. Bank for International Settlements

Bell, E. A., Boehnke, P., Harrison, T. M., Mao, W. L. (2015). Potentially biogenic carbon preserved in a 4.1 billion-year-old zircon. Proceedings of the National Academy of Sciences, 112(47), 14518–14521.

Bennett, C. H. (1985). Dissipation Information Complexity Organization. In Santa Fe Institute Studies in the Sciences of Complexity (pages 215–231). Addison-Wesley Publishing Co.

© The Author(s) 2020
M. Alemi, *The Amazing Journey of Reason*, SpringerBriefs in Computer Science, https://doi.org/10.1007/978-3-030-25962-4

Berners-Lee, T., Fischetti, M. (1999). Weaving the Web: The original design and ultimate destiny of the World Wide Web by its inventor. DIANE Publishing Company.

Berners-Lee, T., Hendler, J., Lassila, O. (2001). The Semantic Web. A new form of Web content that is meaningful to computers will unleash a revolution of new possibilities. Scientific American, 284(5), 3.

Boesch, C., Boesch, H. (1990). Tool Use and Tool Making in Wild Chimpanzees. Folia Primatologica, 54, 86–99.

Borrell, V., Calegari, F. (2014). Mechanisms of brain evolution: Regulation of neural progenitor cell diversity and cell cycle length. Neuroscience Research, 86, 14–24.

Bostrom, N. (2003). Ethical issues in advanced artificial intelligence. Science Fiction and Philosophy: From Time Travel to Superintelligence, 277–284.

Botha, R., Knight, C. (Eds.). (2009). The cradle of language (Vol. 12). OUP Oxford.

Breuer, T., Ndoundou-Hockemba, M., Fishlock, V. (2005). First observation of tool use in wild gorillas. PLoS Biology, 3(11), e380.

Brin, S., Page, L. (1998). The anatomy of a large-scale hypertextual web search engine. Computer networks and ISDN systems, 30(1–7), 107–117.

Buchanan, M. (2007). The social atom. Bloomsbury, New York, NY, USA.

Burbidge, E. Margaret; Burbidge, G. R.; Fowler, William A.; Hoyle, F. (1957). "Synthesis of the Elements in Stars". Reviews of Modern Physics. 29 (4): 547–650

Calcott, B., Sterelny, K. (Eds.). (2011). The major evolutionary transitions revisited. The MIT Press.

Casaleggio, D. (2008). Tu sei rete. La Rivoluzione del business, del marketing e della politica attraverso le reti sociali.

Chao, X., Kou, G., Li, T., Peng, Y. (2018). Jie Ke versus AlphaGo: A ranking approach using decision making method for large-scale data with incomplete information. European Journal of Operational Research, 265(1), 239–247.

Chitty, D. (1996). Do lemmings commit suicide?. Oxford University Press.

Christian, D. (2011). Maps of time: An introduction to big history. Univ of California Press.

Coen, E., Coen, J. (2009). The Big Lebowski. Faber & Faber.

Cohen, M. N. (1977). The food crisis in prehistory. Overpopulation and the origins of agriculture.

Cole, L. C. (1954). The Population Consequences of Life History Phenomena. The Quarterly Review of Biology, 29(2), 103–137.

Cootner, P.H., (editor) 1964. The random character of stock market prices. M.I.T. Press

Corballis, M. C. (2015). Did Language Evolve before Speech? In R. Larson, V. Déprez, H. Yamakido (Eds.), The Evolution of Human Language: Biolinguistic Perspectives (Approaches to the Evolution of Language) (pp. 1–17). Cambridge University Press.

Couzin, I. D. (2009). Collective cognition in animal groups. Trends in cognitive sciences, 13(1), 36–43.

Cringely, R. X. (1996). Accidental empires. New York: HarperBusiness.

Darwin, C. (1874). The descent of man and selection in relation to sex. The Descent of Man and Selection in Relation to Sex (2nd ed., Vol. 1). John Murray.

Davey, M. E., O'toole, G. A. (2000). Microbial biofilms: from ecology to molecular genetics. Microbiology and molecular biology reviews, 64(4), 847–867.

Davidson, E. H., Rast, J. P., Oliveri, P., Ransick, A., Calestani, C., Yuh, C. H., ... Otim, O. (2002). A genomic regulatory network for development. science, 295(5560), 1669–1678.

Davies, P. (2000). The fifth miracle: The search for the origin and meaning of life. Simon and Schuster. Edizione italiana: "Da dove viene la vita" Saggi Mondadori (2000).

Dawkins, R. (1976). The selfish gene. Oxford university press. New York.

Dawkins, R. (2013). The Selfish Gene. Journal of Chemical Information and Modelling, 53(9), 1689–1699.

De Finetti, B. (1970). Teoria delle probabilità. Einaudi. English translation: De Finetti, B. (2017). Theory of probability: a critical introductory treatment (Vol. 6). John Wiley & Sons.

De Waal, F. B. M. (1988). The communicative repertoire of captive bonobos, Pan paniscus, compared to that of chimpanzees. Behaviour 106: 183–251

Dean, J., Ghemawat, S. (2004). MapReduce: simplified data processing on large clusters. In Proceedings of the 6th conference on Symposium on Operating Systems Design & Implementation - Volume 6 (OSDI'04), Vol. 6. USENIX Association, Berkeley, CA, USA, 10–10.

Decho, A. W., Visscher, P. T., Ferry, J., Kawaguchi, T., He, L., Przekop, K. M., ... Reid, R. P. (2009). Autoinducers extracted from microbial mats reveal a surprising diversity of N-acylhomoserine lactones (AHLs) and abundance changes that may relate to diel pH. Environmental Microbiology, 11(2), 409–420.

Dehaene, S. (2014). Consciousness and the brain: Deciphering how the brain codes our thoughts. Penguin.

Denbigh, N (1981) How subjective is entropy? Chem. Brit. 17, 168 85

Deneubourg, J. L., Goss, S. (1989). Collective patterns and decision-making. Ethology Ecology & Evolution, 1(4), 295–311.

Diamond, J. (2002). Evolution, consequences and future of plant and animal domestication. Nature, 418 (6898), 700.

Donaldson, L. (2001). The contingency theory of organizations. Sage.

Dorigo, M., Gambardella, L. M. (1997). Ant colony system: A cooperative learning approach to the traveling salesman problem. IEEE Transactions on Evolutionary Computation, 1(1), 53–66.

Dorogovtsev, S. N., Mendes, J. F. (2003). Evolution of networks: From biological nets to the Internet and WWW. Oxford University Press.

Dunbar, R. I. (1992). Neocortex size as a constraint on group size in primates. Journal of human evolution, 22(6), 469–493.

Dunbar, R. I. (1998). The Social Brain Hypothesis. Evolutionary Anthropology, 6(5), 178–190.

Dunbar, R. I. M. (2004). Gossip in Evolutionary Perspective. Review of General Psychology, 8(2), 100–110.

Dunnington, G. W., Gray, J., Dohse, F. E. (2004). Carl Friedrich Gauss: Titan of science. MAA.

Dyson, F. (2004). A meeting with Enrico Fermi How one intuitive physicist rescued a team from fruitless research. Nature, 427(January), 8540.

Dyson, F. (2004). Origins of Life. Cambridge University Press.

Easley, D., Kleinberg, J. (2010). Networks, crowds, and markets: Reasoning about a highly connected world. Cambridge University Press.

Eccles, J. J. (1989). Evolution of the Brain: Creation of the Self. Routledge.

Eco, U., (1962). Opera aperta: Forma e indeterminazione nelle poetiche contemporanee (Vol. 3). Tascabili Bompiani. Translated in: Eco, U., 1989. The Open Work, trans. Anna Cancogni. Hutchinson, Radius (USA, NP).

Edwards, D. (2011). I'm feeling lucky: The confessions of Google employee number 59. HMH.

Einstein, A. (1934). On the method of theoretical physics. Philosophy of science, 1(2), 163–169.

Epigenetics (2013) An Overview of the Molecular Basis of Epigenetics in Epigenetic Regulation in the Nervous System

Erdös, P., Rényi, A. (1960). On the evolution of random graphs. Publ. Math. Inst. Hung. Acad. Sci, 5(1), 17–60.

Everett, D. (2017). How language began: The story of humanity's greatest invention. Profile Books.

Everett, D. L. (2009). Don't sleep, there are snakes: Life and language in the Amazonian jungle. Profile Books.

Fama, E.F., 1965. The behavior of stock-market prices. The Journal of Business, 38(1), pp. 34–105.

Ferrill, A. (2018). The origins of war: From the stone age to Alexander the Great. Routledge.

Fisher, R. (1930). The Genetical Theory of Natural Selection. Genetics, 154, 272.

Ford, M. (2018) Architects of Intelligence: The Truth about AI from the People Building It. Packt Publishing

Gaston, K. J., Blackburn, T. M. (1997). How many birds are there?. Biodiversity & Conservation, 6(4), 615–625.

Gazzaniga, M.S., 2008. Human: The science behind what makes us unique. Harper Perennial

Ghemawat, S., Gobioff, H., Leung, S. T. (2003). The Google file system.

Goodall, J. (1964). Tool-using and aimed throwing in a community of free-living chimpanzees. Nature, 201(4926), 1264.

Gooday, A. J., Bowser, S. S., Bett, B. J., Smith, C. R. (2000). A large testate protist, Gromia sphaerica sp. nov.(Order Filosea), from the bathyal Arabian Sea. Deep Sea Research Part II: Topical Studies in Oceanography, 47(1–2), 55–73.

Goss, S., Deneubourg, J. L. (1988). Autocatalysis as a source of synchronised rhythmical activity in social insects. Insectes Sociaux, 35(3), 310–315.

Gray, J. (2018). Seven Types of Atheism. Penguin UK.

Gregory, G. (1763). A New and complete dictionary of arts and sciences. Collins and Co.

Griffith, B., Loveless, D. J. (Eds.). (2013). The Interdependence of Teaching and Learning. IAP.

Grignetti, M. (1964). A Note on the Entropy of Words in Printed English. Information and Control, 7, 304–306.

Guseva, E., Zuckermann, R. N., Dill, K. A. (2017). Foldamer hypothesis for the growth and sequence differentiation of prebiotic polymers. Proceedings of the National Academy of Sciences, 114(36),

Gustin, S (2012) In Kodak Bankruptcy, Another Casualty of the Digital Revolution. Time Magazine. Website: http://business.time.com/2012/01/20/in-kodak-bankruptcy-another-casualty-of-the-digital-revolution/

Haldane J. B. S. (1932). The causes of evolution. Harper and Brothers, London.

Halevy, A., Norvig, P., Pereira, F. (2009). The unreasonable effectiveness of data. in IEEE Intelligent Systems, vol. 24, no. , pp. 8–12

Hall, O., Rao K. (1999) Photosynthesis. Cambridge University Press

Hamilton, W. D. (1964). The genetical evolution of social behaviour. I. Journal of Theoretical Biology, 7(1), 1–16.

Hammer, B. K., and B. L. Bassler. "Quorum Sensing Controls Biofilm Formation in Vibrio Cholerae." Molecular microbiology 50.1 (2003): 101–4.

Hanczyc, M. M., Fujikawa, S. M., Szostak, J. W. (2003). Experimental models of primitive cellular compartments: encapsulation, growth, and division. Science, 302(5645), 618–622.

Harari, Y. N. (2014). Sapiens: A brief history of humankind. Random House.

Hargittai, I. (2008). The martians of science: Five physicists who changed the twentieth century. Oxford University Press.

Harris, M. (1978). Cannibals and Kings. The Origins of Cultures. Comparative Studies in Society and History. William Collins.

Harris, W. C. (1889) Nature, Vol. 1: A Weekly Journal for the Gentleman Sportsman, Tourist and Naturalist

Haslam, M., Hernandez-Aguilar, A., Ling, V., Carvalho, S., Torre, I. de la, DeStefano, A., et al. (2009). Primate archaeology. Nature, 460, 339–344.

Henderson, R. M., Clark, K. B. (1990). Architectural Innovation: The Reconfiguration of Existing Product Technologies and the Failure of Established Firms. Administrative Science Quarterly, 35(1), 9.

Herrnstein, R. J. (1995). The bell curve debate: History, documents, opinions. R. Jacoby, & N. Glauberman (Eds.). New York Times Books.

Hicks, S., Cavanough, M. C., O'Brien, E. (1962). Effects Of Anoxia On The Developing Cerebral Cortex In The Rat. The American Journal of Pathology, XL(6).

Hinton, G. E., Anderson, J. A. (Eds.). (2014). Parallel models of associative memory: updated edition. Psychology press.

Hinton, G. E., Nowlan, S. J. (1987). How Learning Can Guide Evolution. Complex Systems.

Hoffman, P. (1998). The man who loved only numbers. New York: Hyperion Cop. 1998.

Hofman, M. A. (2014). Evolution of the human brain: when bigger is better. Frontiers in neuro-anatomy, 8, 15.

Hölldobler, B., Wilson, E. (2009). The Superorganism. The Beauty Elegance and Strangeness of Insect Societies. W. W. Norton Company

Hollenstein, N., Aepli, N. (2014). Compilation of a Swiss German dialect corpus and its application to pos tagging. In Proceedings of the First Workshop on Applying NLP Tools to Similar Languages, Varieties and Dialects (pp. 85–94).

Hopfield, J. (1982). Neural networks and physical systems with emergent collective computational abilities. Proceedings of the National Academy of Sciences, 79(8), 2554–2558.

Hopfield, J. J. (1982). Neural networks and physical systems with emergent collective computational abilities. Proceedings of the National Academy of Sciences, 79(8), 2554–2558.

Høyrup, J. (2009). State, "Justice", Scribal Culture and Mathematics in Ancient Mesopotamia. Sarton Chair Lecture. Sartoniana, 22, 13–45.

Huxley, T. H., Huxley, L. (1900). Life and Letters of Thomas Henry Huxley (Vol. 1). London: Macmillan.

Il Sole 24 Ore, June 22, 2018. "Salvini contro i vaccini: «Dieci sono inutili e dannosi». Poi ringrazia i due paladini no-vax", Il Sole 24 Ore, June 22, 2018. https://www.ilsole24ore.com/art/notizie/2018-06-22/salvini-contro-vaccini-dieci-sono-inutili-e-dannosi-poi-ringrazia-due-paladini-no-vax-132352.shtml

Isler, K., Van Schaik, C. P. (2008). Why are there so few smart mammals (but so many smart birds)?. Biology Letters, 5(1), 125–129.

Isler, K., Van Schaik, C. P. (2012). How our ancestors broke through the gray ceiling: Comparative evidence for cooperative breeding in early homo. Current Anthropology, 53(S6), S453–S465.

Jablonka, E., Lamb, M. J. (2006). The evolution of information in the major transitions. Journal of Theoretical Biology, 239(2), 236–246.

Jablonka, E., Raz, G. (2009). Transgenerational Epigenetic Inheritance: Prevalence, Mechanisms, and Implications for the Study of Heredity and Evolution. The Quarterly Review of Biology, 84(2), 131–176.

Jackson, D. E., Ratnieks, F. L. W. (2006). Communication in ants. Current Biology, 16(15), 570–574.

Jarrell, T. A., Wang, Y. Y., Bloniarz, A. E., Brittin, C. A., Xu, M., Thomson, J. N., Emmons, S. W. (2012). The Connectome of a Decision Making Neural Network. Science, in press(6093), 437–444.

Johnson, B., Sheung Kwan Lam, S. (2010). Self-organization, Natural Selection, and Evolution: Cellular Hardware and Genetic Software. BioScience, 60(11).

Jones, C. B. (2012). Robustness, Plasticity, and Evolvability in Mammals (SpringerBriefs in Evolutionary Biology). Springer.

Kalan, A. K., Rainey, H. J. (2009). Hand-clapping as a communicative gesture by wild female swamp gorillas. Primates, 50(3), 273–275.

Kandel, E.R., Schwartz, J.H. and Jessell, T.M. eds., (2000). Principles of neural science. New York: McGraw-hill.

Kaufmann, E. (2016). It's NOT the economy, stupid: Brexit as a story of personal values. British Politics and Policy at LSE.

Kelvin, W. T. B. (1891). Popular lectures and addresses. Macmillan and Company.

Kemps, T.. (2005). The Origin and Evolution of Mammals. Oxford University Press.

Kolmogorov, A. (1965). Three approaches to the quantitative definition of information. Problemy Peredachi Informatsii, 1(1), 3–11.

Kondepudi, D., and Prigogine, I. (1998). Modern Thermodynamics. From heat Engines to Dissipative Structures (2nd ed.). John Wiley & Sons.

Koshland, D. E. (1958). Application of a theory of enzyme specificity to protein synthesis. Proceedings of the National Academy of Sciences, 44(2), 98–104.

Kuhn, T. S. (1962). The structure of scientific revolutions. University of Chicago Press.

L'Espresso September 3, 2018. "Rocco Casalino, Luca Morisi e gli altri: ecco chi gestisce il 'ministero della Propaganda'", L'Espresso September 3, 2018. http://espresso.repubblica.it/inchieste/2018/09/03/news/rocco-casalino-luca-morisi-e-gli-altri-chi-gestisce-ministero-propaganda-1.326445

La Repubblica, January 23, 2019. "Vaccini, è bufera sul convegno no-vax organizzato dal M5s nella sala stampa di Montecitorio", La Repubblica, January 23, 2019. https://www.repubblica.it/politica/2019/01/23/news/camera_bufera_su_convegno_no-vax_m5s-217269581/

Ladevèze, S., de Muizon, C., Beck, R. M., Germain, D., Cespedes-Paz, R. (2011). Earliest evidence of mammalian social behaviour in the basal Tertiary of Bolivia. Nature, 474(7349), 83.

Lahr, M. M., Rivera, F., Power, R. K., Mounier, A., Copsey, B., et al. (2016). Inter-group violence among early Holocene hunter-gatherers of West Turkana, Kenya. Nature, 529(7586), 394.

Lake, J. A. (2011). Lynn Margulis (1938–2011). Nature

Landauer, R. (1961). Irreversibility and Heat Generation in the Computing Process. IBM Journal, (July), 183–191.

Leakey, L. S., Tobias, P. V., Napier, J. R. (1964). A new species of the genus Homo from Olduvai Gorge.

Lebowitz, F. (1982). Metropolitan life. Fawcett.

LeCun, Y. (1985). Une procédure d'apprentissage pour réseau à seuil asymétrique. Proceedings of Cognitiva 85, 599–604.

LeCun, Y., Bengio, Y., Hinton, G. (2015). Deep learning. Nature, 521(7553), 436–444.

Lerner, F. (2009). The story of libraries: From the invention of writing to the computer age. Bloomsbury Publishing.

Levi-Montalcini, R. (1987). Elogio dell'imperfezione. Garzanti Ed., Milano.

Lewis, R. (1960). The Evolution Man, or, How I Ate My Father. Pantheon. Translated in Italian as "The Greatest Ape of the Pleistocene" or "Il più grande uomo scimmia del Pleistocene". Adelphi

Li, W., Ma, L., Yang, G., Gan, W. (2017). REM sleep selectively prunes and maintains new synapses in development and learning. Nature Neuroscience, 4(11), 427–437.

Li, X., Harbottle, G., Zhang, J., Wang, C. (2003). The earliest writing? Sign use in the seventh millennium BC at Jiahu, Henan Province, China. Antiquity, 77(295), 31–44.

Li, Y. H., Tian, X. (2012). Quorum sensing and bacterial social interactions in biofilms. Sensors, 12(3), 2519–2538.

Lieberman, D. (1998). Sphenoid shortening and the evolution of modern human cranial shape. Nature, 393(6681), 158.

Lieberman, D. (2014). The story of the human body: evolution, health, and disease. Vintage.

Lincoln, M., Wasser, A. (2013). Spontaneous creation of the Universe Ex Nihilo. Physics of the Dark Universe, 2(4), 195–199.

Linschitz, H. (1953). The information content of a bacterial cell, in H. Questler, ed., Information theory in biology, 251– 262, Urbana, Univ. of Illinois Press.

Lorenz, E. N. (1963). Deterministic nonperiodic flow. Journal of the atmospheric sciences, 20(2), 130–141.

Lorenz, E. N. (1972). Predictability: Does the Flap of a Butterfly's Wings in Brazil Set off a Tornado in Texas?. Talk to the 139th Meeting of the American Association for the Advancement of Science, Washington DC, on December 29, 1972. Reproduced in: Abraham, R., Ueda, Y. (2000). The chaos avant-garde: Memories of the early days of chaos theory (Vol. 39). World scientific.

Loughry, W. J., Prodöhl, P. A, McDonough, C. M., Avise, J. C. (1998). Polyembrony in armadillos. American Scientist, 86(3), 274–279.

Lovelace, C. (1964). Practical theory of three-particle states. I. Nonrelativistic. Physical Review, 135(5B), B1225.

Luncz, L. V., Wittig, R. M., Boesch, C. (2015). Primate archaeology reveals cultural transmission in wild chimpanzees (pan troglodytes verus). Philosophical Transactions of the Royal Society B: Biological Sciences, 370(1682).

Lyon, P. (2007). From quorum to cooperation: lessons from bacterial sociality for evolutionary theory. Studies in History and Philosophy of Science Part C: Studies in History and Philosophy of Biological and Biomedical Sciences, 38(4), 820–833.

Lyon, P. (2015). The cognitive cell: bacterial behavior reconsidered. Frontiers in microbiology, 6, 264.

Lyons, N. A., Kolter, R. (2015). On The Evolution of Bacterial Multicellularity. Current Opinion in Microbiology, 24, 21–28.

Macom, T. E., Porter, S. D. (1995). Food and energy requirements of laboratory fire ant colonies (Hymenoptera: Formicidae). Environmental Entomology, 24(2), 387–391.

Mallory, J. P., Adams, D. Q. (Eds.). (1997). Encyclopedia of Indo-European Culture. Taylor & Francis.

Mandelbrot, B. (1953). An informational theory of the statistical structure of language. Communication Theory, (2), 486–502.

Mandelbrot, B. 1963. The Variation of certain speculative prices

Margulis, L. (1970). Origin of eukaryotic cells: Evidence and research implications for a theory of the origin and evolution of microbial, plant and animal cells on the precambrian Earth. Yale University Press.

Margulis, L. (2008). Symbiotic planet: a new look at evolution. Basic Books.

Margulis, L., Sagan, D. (1997). Microcosmos: Four billion years of microbial evolution. Univ of California Press.

Mari, X., Kiørboe, T. (1996). Abundance, size distribution and bacterial colonization of transparent exopolymeric particles (TEP) during spring in the Kattegat. Journal of Plankton Research, 18(6), 969–986.

Marlow, H., Arendt, D. (2014). Evolution: Ctenophore genomes and the origin of neurons. Current Biology, 24(16), R757–R761.

Marshall, W. F., Young, K. D., Swaffer, M., Wood, E., Nurse, P., Kimura, A., et al. (2012). What determines cell size?. BMC biology, 10(1), 101.

Martin, R. (1996). Scaling of the Mammalian Brain: the Maternal Energy Hypothesis. Physiology, 11(4), 149–156.

Maxwell, J. C. (1871). Theory of Heat. Longmans, Green and Co.

Maynard Smith, J. (1998). Evolutionary Genetics. New York, NY: Oxford University Press.

McBrearty, S., Jablonski, N. G. (2005). First fossil chimpanzee. Nature, 437(7055), 105.

McGrew, W. C. (1998). Culture in Nonhuman Primates? Annual Review of Anthropology, 27(1), 301–328.

Mendelsohn, I. (1946). Slavery in the Ancient near East. The Biblical Archaeologist, 9(4), 74–88.

Miller, G. F. (1998). How mate choice shaped human nature: A review of sexual selection and human evolution. Handbook of evolutionary psychology: Ideas, issues, and applications, 87–129.

Mises, L. von. (1951). Socialism An Economic and Sociobiological Analysis. New Haven Yale University Press.

Moroz, L. L., Kocot, K. M., Citarella, M. R., Dosung, S., Norekian, T. P., Povolotskaya, I. S., ... Kohn, A. B. (2014). The ctenophore genome and the evolutionary origins of neural systems. Nature, 510(7503), 109–114.

Mowshowitz, A., Mitsou, V. (2009). Entropy, orbits and spectra of graphs. Analysis of Complex Networks: From Biology to Linguistics, Wiley-VCH.

Nätt, D., Lindqvist, N., Stranneheim, H., Lundeberg, J., Torjesen, P. A., Jensen, P. (2009). Inheritance of acquired behaviour adaptations and brain gene expression in chickens. PLoS ONE, 4(7).

Nelson, E. J., Harris, J. B., Morris Jr, J. G., Calderwood, S. B., Camilli, A. (2009). Cholera transmission: the host, pathogen and bacteriophage dynamic. Nature Reviews Microbiology, 7(10), 693.

Neubauer, S., Hublin, J. J., Gunz, P. (2018). The evolution of modern human brain shape. Science Advances, 4(1).

Newman, M.E., 2005. Power laws, Pareto distributions and Zipf's law. Contemporary physics, 46(5), pp. 323–351.

Ng, W. L., Bassler, B. L. (2009). Bacterial quorum-sensing network architectures. Annual review of genetics, 43, 197–222.

Niven, J. E., Farris, S. M. (2012). Miniaturization of nervous systems and neurons. Current Biology, 22(9), 323–329

Noble, D. (2010). Letter from Lamarck. Physiology News, 78, 31.

Noble, D. (2015). Evolution beyond neo-Darwinism: a new conceptual framework. Journal of Experimental Biology, 218(1), 7–13.

Norberg, J. (2016). Progress: Ten reasons to look forward to the future. Oneworld Publications.

Nowak, M., Tarnita, C., Wilson, E. (2012). The Evolution of Eusociality. Nature, 466(7310), 1057–1062.

Oparin, A. I. (1957). The Origin of Life on the Earth, 3rd ed., trans. Ann Synge. Edinburgh: Oliver and Boyd.

Pais, A. (1982). Subtle is the Lord: The Science and the Life of Albert Einstein: The Science and the Life of Albert Einstein. Oxford University Press, USA.

Pareto, V. (1906). Manuale di Economia Politica, published by "Società Editrice Libraria"

Pareto, V. (1967). La courbe de la repartition de la richesse, in "Oeuvres complètes: Tome 3, Ecrits sur la courbe de la répartition de la richesse", edited by Giovanni Busino, Librerie Droz, Geneve

Parker, D. B. (1985). Learning Logic Technical Report TR-47. Center of Computational Research in Economics and Management Science, Massachusetts Institute of Technology, Cambridge, MA.

Perelman, G., (2002). The entropy formula for the Ricci flow and its geometric applications. arXiv preprint math/0211159.

Peretó, J., Bada, J. L., Lazcano, A. (2009). Charles Darwin and the origin of life. Origins of life and evolution of biospheres, 39(5), 395–406.

Petronis, A. (2010). Epigenetics as a unifying principle in the aetiology of complex traits and diseases. Nature, 465(7299), 721–727.

Pinker, S. (2003). The language instinct: How the mind creates language. Penguin UK.

Pinker, S. (2011). The better angels of our nature: A history of violence and humanity. Penguin.

Planck, M. (1901). On the law of distribution of energy in the normal spectrum. Annalen der physik, 4(553), 1.

Plato, B. (1914). Plato (Vol. 8). London: Harvard University Press.

Poinar, G. O. (2011). The evolutionary history of nematodes: as revealed in stone, amber and mummies (Vol. 9). Brill.

Popper, K. (1934). Logik der Forschung. Verlag von Julius Springer, Vienna, Austria reprinted in Popper, K. (1959) The logic of scientific discovery. Hutchinson & Co.

Quattrone, G., Proserpio, D., Quercia, D., Capra, L., Musolesi, M. (2016). Who benefits from the sharing economy of Airbnb?. In Proceedings of the 25th international conference on world wide web (pp. 1385–1394). International World Wide Web Conferences Steering Committee

Radnitzky, G., Andersson, G. (Eds.). (1979). The structure and development of science (Vol. 136). Springer Science & Business Media. trad. it. Feyerabend, P. K., Corvi, R. (1989). Dialogo sul metodo. Laterza.

Ratzel, F. (1901). Der lebensraum: Eine biogeographische studie. H. Laupp.

Reader, J. (1998). Africa: A Biography of the Continent. Penguin UK.

Richerson, P. J., Boyd, R., Bettinger, R. L. (2001). Was Agriculture Impossible during the Pleistocene but Mandatory during the Holocene? A Climate Change Hypothesis. Society for American Archaeology, 66(3), 387–411.

Rickard, S. J. (2016). Populism and the Brexit vote. Comparative Politics Newsletter, 26, 120–22.

Robertson, M. P., Joyce, G. F. (2010). The origins of the RNA world. Cold Spring Harbor perspectives in biology, a003608.

Rosenblatt, F. (1958). The perceptron: a probabilistic model for information storage and organization in the brain. Psychological review, 65(6), 386.

Rousseau, J-J (1762). Du contrat social.

Rovelli C. (2015) Relative Information at the Foundation of Physics. In: Aguirre A., Foster B., Merali Z. (editors) It From Bit or Bit From It?. The Frontiers Collection. Springer, Cham

Rubin, D. C. (1995). Memory in oral traditions: The cognitive psychology of epic, ballads, and counting-out rhymes. Oxford University Press

Rumelhart, D. E., Hinton, G. E., Williams, R. J. (1988). Learning representations by back-propagating errors. Cognitive modeling, 5(3)

Russell, B. (1971). Mysticism and Logic, and Other Essays. Barnes & Noble.

Schaller, G. B. (2009). The Serengeti lion: a study of predator-prey relations. University of Chicago Press.

Schrödinger, E. (1944). What is life? The physical aspect of the living cell and mind. Cambridge University Press.

Schultes, R. E., Raffauf, R. F. (1990). The healing forest: medicinal and toxic plants of the Northwest Amazonia. Dioscorides Press.

Schwartz, David N. (2017) The Last Man who Knew Everything: The Life and Times of Enrico Fermi, Father of the Nuclear Age. Hachette UK

Shannon, C. E. (1948). A mathematical theory of communication. Bell system technical journal, 27(3), 379–423.

Shannon, C. E. (1951). Prediction and Entropy of Printed English. Bell Labs Technical Journal, 30(1), 50–64.

Shannon, C. E. (1956). The bandwagon. IRE Transactions on Information Theory, 2(1), 3.

Shapiro, J. A. (1988). Bacteria as multicellular organisms. Scientific American, 258(6), 82–89.

Shen, X. X., Hittinger, C. T., Rokas, A. (2017). Contentious relationships in phylogenomic studies can be driven by a handful of genes. Nature ecology & evolution, 1(5), 0126.

Shih, W (2016) The Real Lessons From Kodak's Decline MIT Sloan Management Review, Summer 2016 Issue. https://sloanreview.mit.edu/article/the-real-lessons-from-kodaks-decline/

Simon, H. A. (1955). On a Class of Skew Distribution Functions. Biometrika, 42(3–4), 425–440.

Simon, H. A. (1962). The architecture of complexity. Proceedings of the American Philosophical Society, 106(6), 467–482.

Small, M., Tse, C. K. (2012). Predicting the outcome of roulette. Chaos: an interdisciplinary journal of nonlinear science, 22(3), 033150.

Smil, V. (2008). Energy in nature and society: general energetics of complex systems. MIT press.

Smith, A. H. (1984). Kutenai Indian Subsistence and Settlement Patterns. Washington State Univ Pullman Center For Northwest Anthropology.

Smith, J. M. (1987). When learning guides evolution. Nature, 329(6142), 761–762.

Smith, J. M., Szathmary, E. (1995). The Major Transitions of Evolution. Evolution, 49(6), 1302.

Sussman, R. W. (2017). The origins and nature of sociality. Routledge.

Swami, V., Barron, D., Weis, L., Furnham, A. (2018). To Brexit or not to Brexit: The roles of Islamophobia, conspiracist beliefs, and integrated threat in voting intentions for the United Kingdom European Union membership referendum. British Journal of Psychology, 109(1), 156–179.

Szathmáry, E. (2001). The origin of the human language faculty: the language amoeba hypothesis. Trends in Linguistics Studies and Monographs, 133, 41–54.

T. Berners-Lee, J. Hendler, and O. Lassila, (17 May 2001). The Semantic Web. Scientific American.

Tabak, J. (2004). Mathematics and the Laws of Nature. Developing the Language of Science. Library of Congress Cataloging-in-Publication Data.

Taglialatela, J. P., Savage-Rumbaugh, S., Baker, L. A. (2003). Vocal Production by a Language-Competent Pan paniscus. International Journal of Primatology, 24(1).

Taleb, N. N., 2007 The Black Swan: The Impact of the Highly Improbable, Random House

Testart, A., Forbis, R. G., Hayden, B., Ingold, T., Perlman, S. M., Pokotylo, D. L., Rowley-Conwy, P., Stuart, D. E. (1982). The significance of food storage among hunter-gatherers: Residence patterns, population densities, and social inequalities. Current anthropology, 23(5), 523–537.

Thorp, E. O. (1998). The invention of the first wearable computer. In Digest of Papers. Second International Symposium on Wearable Computers (Cat. No. 98EX215) (pp. 4–8). IEEE.

Tipografia Forense. (1862). Manuale dei dilettanti della caccia tanto col fucile che colle reti. Roma, Tipografia Forense

Travers, J., Milgram, S. (1967). The small world problem. Psychology Today, 1(1), 61–67.

Travers, J., Milgram, S. (1969). An exploratory study of the small world problem. Sociometry, 32, 425–43.

Tröhler, D. (2005). Langue as homeland: The Genevan reception of pragmatism. In Inventing the Modern Self and John Dewey (pp. 61–83). Palgrave Macmillan, New York.

Tryon, E. P. (1973). Is the universe a vacuum fluctuation?. Nature, 246(5433), 396.

Turing, A. (1950), 'Computing Machinery and Intelligence', Mind 59(236), pp. 433–460.

Turnbull, C. (1961). The forest people. Random House.

van Schaik, C. P., Ancrenaz, M., Borgen, G., Galdikas, B. M. F., Knott, C. D., Singleton, I., et al. (2003). Orangutan cultures and the comparative study of culture. Science, 102(2003), 214.

Varshney, L. R., Chen, B. L., Paniagua, E., Hall, D. H., Chklovskii, D. B. (2011). Structural properties of the Caenorhabditis elegans neuronal network. PLoS computational biology, 7(2), e1001066.

Vilenkin, A. (1982). Creation of universes from nothing. Physics Letters B, 117(1–2), 25–28.

Von Neumann, J. (1956). Probabilistic logics and the synthesis of reliable organisms from unreliable components. Automata studies, 34:

Waal, F. B. M. de. (1996). Good Natured. Harvard University Press.

Watson, D. F., Berlind, A. A., Zentner, A. R. (2011). A cosmic coincidence: The power-law galaxy correlation function. The Astrophysical Journal, 738(1), 22.

Watson, R. A., Szathmáry, E. (2016). How Can Evolution Learn ? Trends in Ecology & Evolution, 31(2), 147–157.

Watson, R. A., Buckley, C. L., Mills, R., Davies, A. (2010). Associative memory in gene regulation networks. Artificial Life XII. Proceedings of the 12th International Conference on the Synthesis and Simulation of Living Systems, 194–202.

Watts, D. J., Strogatz, S. H. (1998). Collective dynamics of 'small-world'networks. Nature, 393(6684), 440.

Weaver, W. (1949). The mathematics of communication. Scientific American, 181(1), 11–15.

Wengrow, D., Graeber, D. (2015). Farewell to the "childhood of man": Ritual, seasonality, and the origins of inequality. Journal of the Royal Anthropological Institute, 21(3), 597–619.

Werbos, P. (1974). Beyond Regression: New Tools for Prediction and Analysis in the Behavioral Sciences. Ph. D. dissertation, Harvard University.

Wheeler, J. A. (1990). Information, physics, quantum: The search for links. Complexity, entropy, and the physics of information, 8.

Wheeler, P., Aiello, L. (1995). The expensive tissue hypothesis. Current Anthropology, 36(2), 199–221.

White, F. J., Waller, M., Boose, K., Merrill, M. Y., Wood, K. D. (2015). Function of loud calls in wild bonobos. Journal of Anthropological Sciences, 93, 89–101.

White, J. G., Southgate, E., Thomson, J. N., Brenner, S. (1986). The structure of the nervous system of the nematode Caenorhabditis elegans. Philos Trans R Soc Lond B Biol Sci, 314(1165), 1–340.

White, T. (2012). Hadoop: The definitive guide. O'Reilly Media, Inc.

Why Technology Favors Tyranny, The Atlantic, October 2018, https://www.theatlantic.com/magazine/archive/2018/10/yuval-noah-harari-technology-tyranny/568330/

Wigner, E. (1960). The Unreasonable Effectiveness of Mathematics in the Natural. Communication in Pure and Applied Mathematics, 13 (I).

Wöhler, F. (1828). Über künstliche bildung des harnstoffs. Annalen der Physik, 88(2), 253–256.

Wolchover, N. (2014). A new physics theory of life. Scientific American.

Wolchover, N. A new theory of the evolution of life. https://www.quantamagazine.org/a-new-thermodynamics-theory-of-the-origin-of-life-20140122/.

Woods, W. A. (1970). Transition network grammars for natural language analysis. Communications of the ACM, 13(10), 591–606.

Wrangham, Richard. Catching fire: how cooking made us human. Basic Books, 2009.

Yockey, H. P. (1977). A calculation of the probability of spontaneous biogenesis by information theory. Journal of Theoretical Biology, 67, 377–398.

Yogeesha, C. B., Pujeri, R. V. (2014). A Comparative Study of Geometric Hopfield Network and Ant Colony Algorithm to Solve Travelling Salesperson Problem. International Journal of Advanced Computer Research, 4(3), 843.

Yoshua Bengio in Ford, M. (2018) Architects of Intelligence: The Truth about AI from the People Building It. Packt Publishing

Zerjal, T., Xue, Y., Bertorelle, G., Wells, R. S., Bao, W., Zhu, S., ... Tyler-smith, C. (2003). The Genetic Legacy of the Mongols. The American Journal of Human Genetics, 72(3), 717–721.

Zhang, G., Zhang, F., Ding, G., Li, J., Guo, X., Zhu, J., Dong, X. (2012). Acyl homoserine lactone-based quorum sensing in a methanogenic archaeon. ISME Journal, 6(7), 1336–1344.

Zhao, H. W., Zhou, D., Nizet, V., Haddad, G. G. (2010). Experimental selection for Drosophila survival in extremely high O2 environments. PLoS ONE, 5(7), 1–8.

Ziolkowski, J. M. (2011). De laude scriptorum manualium and De laude editorum: From Script to Print, From Print to Bytes. Ars edendi Lecture Series, 1, 25–58.

Zipf, G. K. (1932). Selected Studies of the Principle of Relative Frequency in Language. Language. Oxford University Press.

Printed in the United States
By Bookmasters